# QUALITY ASSURANCE FOR
# THE ANALYTICAL CHEMISTRY LABORATORY

# QUALITY ASSURANCE FOR
# THE ANALYTICAL CHEMISTRY LABORATORY

D. Brynn Hibbert

UNIVERSITY PRESS

2007

# OXFORD
**UNIVERSITY PRESS**

Oxford University Press, Inc., publishes works that further
Oxford University's objective of excellence
in research, scholarship, and education.

Oxford   New York
Auckland   Cape Town   Dar es Salaam   Hong Kong   Karachi
Kuala Lumpur   Madrid   Melbourne   Mexico City   Nairobi
New Delhi   Shanghai   Taipei   Toronto

With offices in
Argentina   Austria   Brazil   Chile   Czech Republic   France   Greece
Guatemala   Hungary   Italy   Japan   Poland   Portugal   Singapore
South Korea   Switzerland   Thailand   Turkey   Ukraine   Vietnam

Copyright © 2007 by Oxford University Press, Inc.

Published by Oxford University Press, Inc.
198 Madison Avenue, New York, New York 10016

www.oup.com

Oxford is a registered trademark of Oxford University Press

Library of Congress Cataloging-in-Publication Data
Hibbert, D. B. (D. Brynn), 1951–
Quality assurance for the analytical chemistry laboratory / D. Brynn Hibbert.
p.  cm.
Includes bibliographical references.
ISBN 978-0-19-516212-7; 978-0-19-516213-4 (pbk.)
1. Chemical laboratories—Quality control.   2. Chemistry, Analytic—Quality control.
3. Chemistry, Analytic—Technique.   4. Chemometrics.   I. Title.
QD75.4.Q34H53 2006
542—dc22     2006014548

1   3   5   7   9   8   6   4   2

Printed in the United States of America
on acid-free paper

$54.50

*This book is dedicated to my friends and colleagues on IUPAC project 2001-010-3-500, "Metrological Traceability of Chemical Measurement Results"*

Paul De Bièvre, René Dybkaer, and Alĕs Fajgelj

# Preface

Analytical chemistry impacts on every aspect of modern life. The food and drink we consume is tested for chemical residues and appropriate nutritional content by analytical chemists. Our health is monitored by chemical tests (e.g. cholesterol, glucose), and international trade is underpinned by measurements of what is being traded (e.g. minerals, petroleum). Courts rely more and more on forensic evidence provided by chemistry (e.g. DNA, gun-shot residues), and the war on terrorism has caused new research into detection of explosives and their components. Every chemical measurement must deliver a result that is sufficiently accurate to allow the user to make appropriate decisions; it must be fit for purpose.

The discipline of analytical chemistry is wide and catholic. It is often difficult for a food chemist to understand the purist concerns of a process control chemist in a pharmaceutical company. The former deals with a complex and variable matrix with many standard analytical methods prescribed by Codex Alimentarius, for which comparability is achieved by strict adherence to the method, and the concept of a "true" result is of passing interest. Pharmaceuticals, in contrast, have a well-defined matrix, the excipients, and a well-defined analyte (the active) at a concentration that is, in theory, already known. A 100-mg tablet of aspirin, for example, is likely to contain close to 100 mg aspirin, and the analytical methods can be set up on that premise. Some analytical methods are more stable than others, and thus the need to check calibrations is less pressing. Recovery is an issue for many analyses of environmental samples, as is speciation. Any analysis that must be compared to a regulatory limit risks challenge if a proper measurement uncertainty has not been reported. When any measurement is scrutinized in a court of law, the analyst must be able to defend the result and show that it has been done properly.

Every chemical laboratory, working in whatever field of analysis, is aware of the need for quality assurance of its results. The impetus for this book is to bring together modern thinking on how this might be achieved. It is more than a text book that just offers recipes; in it I have tried to discuss how different actions impact on the analyst's ability to deliver a quality result. The quality manager always has a choice, and within a limited budget needs to make effective decisions. This book will help achieve that goal.

After a general introduction in which I discuss the heart of a chemical measurement and introduce commonly accepted views of quality, some basic statistical tools are briefly described in chapter 2. (My book on data analysis for analytical chemistry [Hibbert and Gooding 2005] will fill in some gaps and perhaps remind you of some of the statistics you were taught in your analytical courses.) Chapter 3 covers experimental design; this chapter is a must read if you ever have to optimize anything. In chapter 4, I present general QC tools, including control charts and other graphical help mates. Quality is often regulated by accreditation to international standards (chapter 9), which might involve participation in interlaboratory studies (chapter 5). Fundamental properties of any measurement result are measurement uncertainty (chapter 6) and metrological traceability (chapter 7). All methods must be validated, whether done in house or by a collaborative study (chapter 8). Each laboratory needs to be able to demonstrate that it can carry out a particular analysis to achieve targets for precision (i.e., it must verify the methods it uses).

There are some existing texts that cover the material in this book, but I have tried to take a holistic view of quality assurance at a level that interested and competent laboratory scientists might learn from. I am continually surprised that methods to achieve quality, whether they consist of calculating a measurement uncertainty, documenting metrological traceability, or the proper use of a certified reference material, are still the subject of intense academic debate. As such, this book runs the risk of being quickly out of date. To avoid this, I have flagged areas that are in a state of flux, and I believe the principles behind the material presented in this book will stand the test of time.

Many quality assurance managers, particularly for field laboratories, have learned their skills on the job. Very few tertiary courses exist to help quality assurance managers, but assiduous searching of the Internet, subscription to journals such as *Accreditation and Quality Assurance*, and participation in the activities of professional organizations allow analysts to build their expertise. I hope that this book will fill in some gaps for such quality assurance personnel and that it will give students and new professionals a head start.

Finally, I am not a guru. Please read this text with the same critical eye that you lend to all your professional work. I have tried to give practical advice and ways of achieving some of the more common goals of quality in analytical chemistry. I hope you will find useful recipes to follow. Have fun!

## Reference

Hibbert, D B and Gooding, J J (2005), *Data Analysis for Chemistry: An Introductory Guide for Students and Laboratory Scientists* (New York: Oxford University Press).

# Acknowledgments

Some of the material for this book comes from a graduate course, "Quality assurance in chemical laboratories," that I have taught for a number of years at the University of New South Wales in Sydney. I am indebted to the many students who have given excellent feedback and hope I have distilled their communal wisdom with appropriate care. I also extend many thanks to my co-teachers, Tareq Saed Al-Deen, Jianfeng Li, and Diako Ebrahimi. Thanks also to my present PhD student Greg O'Donnell for his insights into the treatment of bias.

The community of analytical chemists in Australia is a small one. I occupy the longest established (indeed perhaps the only) chair of analytical chemistry in the country and therefore have been fortunate to participate in many aspects of the nation's analytical and metrological infrastructure. I thank my colleagues at NATA (National Association of Testing Authorities, the world's first accreditation body), particularly Maree Stuart, Regina Robertson, Alan Squirrell, John Widdowson, and Graham Roberts. Alan, sometime chair of CITAC, and Regina were very free with their advice in the early days of the course. I also thank Glenn Barry, who was working on an Australian Standard for soil analysis, for making available the soil data used to illustrate homogeneity of variance in chapter 2. Until Australia's metrological infrastructure was brought under the aegis of the National Measurement Institute (NMI), I was involved with legal metrology through my appointment as a commissioner of the National Standards Commission (NSC). I thank the last chair, Doreen Clark, for her excellent work for the

NSC and analytical chemistry and education in general. I also acknowledge the skill and professionalism of Judith Bennett, Grahame Harvey, Marian Haire, and Yen Heng, and thank them for their help in explaining the field of legal metrology to a newcomer.

The National Analytical Reference Laboratory, now part of the NMI, was set up by Bernard King and then taken forward by Laurie Bezley. I am fortunate in chairing a committee that scrutinizes pure reference materials produced by the NMI and have worked fruitfully with organic chemists Steven Westwood and Stephen Davis. Thanks, too, to Lindsey MacKay, Adam Crawley, and many colleagues at NMI.

The Royal Australian Chemical Institute has supported metrology in chemistry through its "Hitchhiker's Guide to Quality Assurance" series of seminars and workshops. These have been excellently organized by Maree Stuart and John Eames and have been well attended by the analytical community. I particularly thank John Eames for allowing me to use his approach for quality control materials in chapter 4.

My greatest thanks go to my three colleagues from the IUPAC project "Metrological Traceability of Measurement Results in Chemistry," to whom this book is dedicated. If I have learned anything about metrology in chemistry, it is from Paul De Bièvre, René Dybkaer, and Aleš Fajgelj.

I thank the editors and production staff at Oxford University Press for efficiently turning my Australian prose into text that can be understood by a wider audience.

Finally, thanks and love to my family, Marian Kernahan, Hannah Hibbert, and Edward Hibbert, for continual support and encouragement. Was it worth it? I think so.

# Contents

# QUALITY ASSURANCE FOR
# THE ANALYTICAL CHEMISTRY LABORATORY

# 1

## Introduction to Quality in
## the Analytical Chemistry Laboratory

### 1.1 Measurement in Chemistry

#### 1.1.1 Defining Measurement

To understand quality of chemical measurements, one needs to understand something about measurement itself. The present edition of the *International Vocabulary of Basic and General Terms in Metrology* (ISO 1993, term 2.1)[1] defines a measurement as a "set of operations having the object of determining a value of a quantity." Quantity is defined as an "attribute of a phenomenon, body or substance that may be distinguished qualitatively and determined quantitatively" (ISO 1993, term 1.1). Typical quantities that a chemist might be interested in are mass (not weight), length, volume, concentration, amount of substance (not number of moles), current, and voltage. A curse of chemistry is that there is only one unit for amount of substance, the mole, and perhaps because "amount of substance" is verbally unwieldy and its contraction "amount" is in common nonscientific usage, the solecism "number of moles" is ubiquitous and has led to general confusion between quantities and units.

The term "measurand," which might be new to some readers, is the quantity intended to be measured, so it is correct to say of a numerical result that it is the value of the measurand. Do not confuse measurand with analyte. A test material is composed of the analyte and the matrix, and so the measurand is physically embodied in the analyte. For example, if the measurand is the mass fraction of dioxin in a sample of pig liver, the dioxin is the analyte and the liver is the matrix. A more rigorous approach of defining a quantity in terms of, System – Component; kind of quantity, has been under discussion in clinical medicine for some time. This concept of specifying a quantity has recently been put on a sound ontological footing by Dybkaer (2004).

A measurement result typically has three components: a number and an uncertainty with appropriate units (which may be 1 and therefore conventionally omitted). For example, an amount concentration of copper might be $3.2 \pm 0.4$ µmol $L^{-1}$. Chapter 6 explains the need to qualify an uncertainty statement to describe what is meant by plus or minus (e.g., a 95% confidence interval), and the measurand must also be clearly defined, including speciation, or isomeric form. Sometimes the measurement is defined by the procedure, such as "pH 8 extractable organics."

### 1.1.2   The Process of Analysis

Analytical chemistry is rarely a simple one-step process. A larger whole is often subsampled, and the portion brought to the laboratory may be further divided and processed as part of the analysis. The process of measurement often compares an unknown quantity with a known quantity. In chemistry the material embodying the known quantity is often presented to the measurement instrument first, in a step called calibration. Because of the complexity of matrices, an analyst is often uncertain whether all the analyte is presented for analysis or whether the instrument correctly responds to it. The measurement of a reference material can establish the recovery or bias of a method, and this can be used to correct initial observations. Figure 1.1 is a schematic of typical steps in an analysis. Additional steps and measurements that are part of the quality control activities are not shown in this figure.

## 1.2  Quality in Analytical Measurements

We live in the age of quality. Quality is measured, analyzed, and discussed. The simplest product and the most trivial service come from quality-assured organizations. Conspicuously embracing quality is the standard of the age. Even university faculty are now subject to "quality audits" of their teaching. Some of these new-found enthusiasms may be more appropriate than others, but I have no doubt that proper attention to quality is vital for analytical chemistry. Analytical measurements affect every facet of our modern, first-world lives. Health, food, forensics, and general trade require measurements that often involve chemical analysis, which must be accurately conducted for informed decisions to be made. A sign of improvement in developing countries is often a nation's ability to measure important aspects of the lives of its citizens, such as cleanliness of water and food.

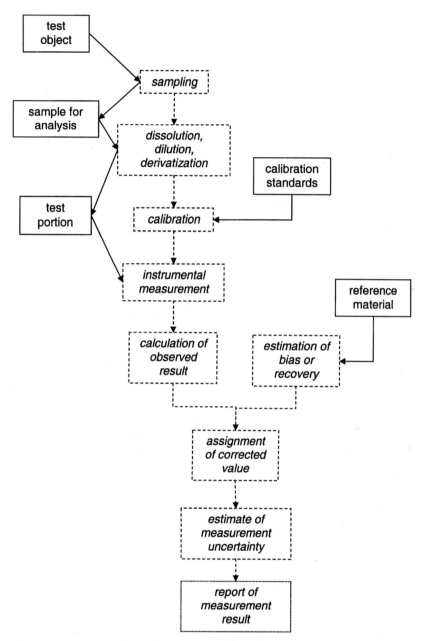

**Figure 1.1.** Steps and materials in an analysis. Procedures are shown in dashed boxes. Quality control materials that are presented to the analytical system are not shown.

### 1.2.1  The Cost of Quality

A well-known saying that can be applied to many areas of life is "if you think quality systems are expensive, look at the cost of not having them." The prevention of a single catastrophic failure in quality that might result in great loss (loss of life, loss of money through lawsuits, loss of business through loss of customer confidence) will pay for a quality system many times over. Of course, prevention of an outcome is more difficult to quantify than the outcome itself, but it can be done. Figure 1.2 is a conceptual graph that plots the cost of quality systems against the cost of failures. The cost of quality, after a setup cost, is a linear function of the activity. The more quality control (QC) samples analyzed, the more QA costs. Failures decrease dramatically with the most rudimentary quality system, and after a while the system is close to optimum performance. (This statement is made with due deference to the continuous-improvement school of total quality management.) The combination of the costs and savings gives a point at which an optimum amount of money is being spent. Remaining at the minimum failure point in the graph requires more work to reduce the point still further (and this is where total quality management [TQM] comes in). It is difficult to give an accurate graph for a real situation. The cost of the quality system can be determined, but the cost of failures is less well known. Most companies do not have the luxury of operating without a quality system simply to quantify the cost of failure.

I preface my lectures on quality assurance in the chemical laboratory by asking the rhetorical question, why bother with quality? The answer is "because it costs a lot to get it wrong." There are many examples of failures in chemical analysis that have led to great material loss, but as a first example here is a success story.

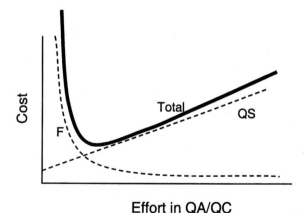

**Figure 1.2.** The cost of quality. F = cost of failure, QS = cost of the quality system. The minimum in the combined graph is the optimum overall cost.

The United States has been monitoring some of its common medical tests by reanalyzing samples using a more accurate method and determining levels of false positives and false negatives. In 1969 the false positive rate on cholesterol tests (concluding a patient has high cholesterol when he or she does not) was 18.5%. By 1994, when presumably modern enzyme methods were being used, the false positive rate was down to 5.5–7.2%, with concomitant savings of $100 million per year. The savings arise from not repeating doctor's visits, not prescribing unnecessary medication, and not adopting costly diets for people who, in fact, do not have a cholesterol problem.

During the same period, NIST (the National Institute of Standards and Technology, formerly the National Bureau of Standards) reported that the cost of nondiagnostic medical tests in the United States at the end of the 1990s was $36 billion, about one-third of the total cost of testing. Not all these tests are chemical, and so not all the retests would have been a result of poor quality in a laboratory, but the figure is very large (U.S. Senate 2001).

In recent years Chile has fallen foul of both the United States (because a grape crop allegedly contained cyanide; De Bievre 1993) and the European Union (because shrimp that contained cadmium below the limit of defensible detection was rejected), and each time Chile suffered losses in the millions of dollars. In a survey of users of analytical chemical results, the Laboratory of the Government Chemist (LGC) in the United Kingdom found that 29% of the respondents to a survey had suffered loss as a result of poor analytical chemistry, and 12% of these claimed "very serious" losses (King 1995).

It was stories such as these, circulating at the end of the twentieth century, that stirred the world of analytical chemistry and have caused analytical chemists to look at how a venerable profession is apparently in such strife.[2]

Even when the analysis is being performed splendidly, the limitation of any measurement due to measurement uncertainty always leads to some doubt about the result. See chapter 6 for an example of uncertainty concerning the amount of weapons-grade plutonium in the world.

### 1.2.2  Definitions of Quality

There is no lack of definitions of quality. Here are some general ones:

- Delivering to a customer a product or service that meets the specification agreed on with the customer, and delivering it on time
- Satisfying customer requirements
- Fitness for purpose
- Getting it right the first time.

The International Organization for Standardization (ISO) definitions of quality are:

- The totality of features and characteristics of a product or service that bear on its ability to satisfy stated or implied needs (ISO 1994)

- Degree to which a set of inherent characteristics fulfils requirements (ISO 2005), where "characteristics" are *distinguishing features*, and "requirements" are *need or expectation that is stated, generally implied, or obligatory.*

Clearly, quality is all about satisfying a customer. Herein lies the first problem of an analytical chemist. When a customer buys a toaster, his or her needs are satisfied if the appliance does indeed toast bread to a reasonable degree, in a reasonable time, and if the toaster does not cause a fire that burns down the kitchen. Many analytical measurements, whether they are made after a visit to the doctor or before a food product is sold, are done without the explicit knowledge or understanding of the consumer. Occasionally, perhaps after a misdiagnosis based on a laboratory analysis, a failure of quality might become apparent, but for the most part results are taken largely on trust. There is often a "middle man," a government department or medical personnel, who is better placed to assess the results, and this is how the public learns of the general concerns over quality. Knowing the requirements of the customer does allow some of the quality parameters to be set. The method must work within a certain concentration range and with a particular limit of detection; the measurement uncertainty must be appropriate to the end user's needs; and the cost and time of delivery of the results must be acceptable.

The assessment of the quality of a result must be drawn from a number of observations of the laboratory, the personnel, the methods used, the nature of the result, and so on. The great leap forward in understanding quality came in the twentieth century when people such as Deming, Shewhart, Ishikawa, and Taguchi formulated principles based on the premise that the quality of a product cannot be controlled until something is measured (Deming 1982; Ishikawa 1985; Roy 2001; Shewhart 1931). Once measurement data are available, statistics can be applied and decisions made concerning the future.

### 1.2.2.1 Quality Systems, Quality Control, and Quality Assurance

The Association of Official Analytical Chemists (AOAC, now AOAC International), uses the following definitions (AOAC International 2006):

*Quality management system*: Management system to direct and control an organization with regard to quality (AOAC International 2006, term 31)

*Quality control*: Part of quality management focused on fulfilling quality requirements (AOAC International 2006, term 29)

*Quality assurance*: Part of quality management focused on providing confidence that quality requirements will be fulfilled (AOAC International 2006, term 28).

A quality system is the overarching, enterprise-level operation concerned with quality. The day-to-day activities designed to monitor the process are the business of quality control (QC), while the oversight of the QC activities belongs to the quality assurance (QA) manager. Some definitions discuss quality in terms of planned activities. Noticing quality, or more likely the lack of it, is not a chance occurrence. Vigilant employees are to be treasured, but a proper quality system has been carefully thought out before a sample is analyzed and entails more than depending on conscientious employees. The way the activities of a quality system might be seen in terms of a measurement in an analytical chemistry laboratory is shown in figure 1.3.

### 1.2.2.2 Qualimetrics

In recent years the term "qualimetrics" has been coined to refer to the use of chemometrics for the purposes of quality control (Massart et al. 1997). It relates particularly to the use of multivariate analysis of process control measurements. Other texts on quality assurance in chemical laboratories include the latest edition of Garfield's book published by AOAC International (Garfield et al. 2000), material published through the Valid Analytical Measurement program by the LGC (Prichard 1995), and books from the Royal Society of Chemistry (Parkany 1993, 1995; Sargent and MacKay 1995). Wenclawiak et al. (2004) have edited a series of Microsoft PowerPoint presentations on aspects of quality assurance.

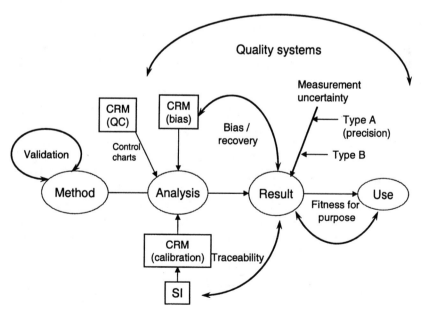

**Figure 1.3.** A schematic of aspects of quality in an analytical measurement.

### 1.2.2.3   Valid Analytical Measurement

The Valid Analytical Measurement (VAM; LGC 2005) program of the LGC (the U.K. National Measurement Institute for chemical measurements) typifies a modern approach to quality in chemical measurements. The program's six principles are a clear exposition of the important aspects of making reliable analytical measurements:

1. Analytical measurements should be made to satisfy an agreed requirement.
2. Analytical measurements should be made using methods and equipment that have been tested to ensure they are fit for purpose.
3. Staff making analytical measurements should be both qualified and competent to undertake the task.
4. There should be a regular independent assessment of the technical performance of a laboratory.
5. Analytical measurements made in one location should be consistent with those elsewhere.
6. Organizations making analytical measurements should have well-defined quality control and quality assurance procedures.

Each of these principles will arise in some guise or other in this book. For example, principle 5 relates to metrological traceability (chapter 7) and measurement uncertainty (chapter 6). These principles will be revisited in the final chapter.

## 1.3 The International System of Measurement

### 1.3.1   The Treaty of the Metre

The French revolution of 1789 gave an opportunity for the new regime under Talleyrand to lay down the basic principles of a universal measurement system. By 1799 the Metre and Kilogram of the Archives, embodiments in platinum of base units from which other units were derived, became legal standards for all measurements in France. The motto of the new metric system, as it was called, was "for all people, for all time." Unfortunately, despite initial support from England and the United States, the new system was confined to France for three quarters of a century. The Treaty of the Metre was not signed until 1875, following an international conference that established the International Bureau of Weights and Measures. Having universally agreed-upon units that would replace the plethora of medieval measures existing in Europe opened possibilities of trade that, for the first time, would allow exchange of goods (and taxes to be levied) on a standardized basis. The original 18 countries that signed the Treaty of the Metre have now become 51, including all the major trading nations, and the ISQ (international system of quantities) of which the SI is the system of units, is the only system that can claim to be worldwide. Table 1.1 lists the base quantities and base units in the SI system, and table 1.2 lists the derived quantities and units.

### 1.3.2 International Metrology

The General Conference on Weights and Measures (CGPM) meets every 4 years and makes additions to, and changes in, the international system of units (SI).[3] A select group of 18 internationally recognized scientists from the treaty nations is the International Committee of Weights and Measures

Table 1.1.  Base quantities and their base units in SI, as determined by the General Conference of Weights and Measures (BIPM 2005)

| Quantity | Unit (symbol) | Definition of unit |
|---|---|---|
| mass | kilogram (kg) | The mass of the international prototype of the kilogram |
| length | meter (m) | The length of the path traveled by light in a vacuum in 1/299,792,458 second |
| time | second (s) | 9,192,631,770 cycles of radiation associated with the transition between the two hyperfine levels of the ground state of the cesium-133 atom |
| thermodynamic temperature | kelvin (K) | 1/273.16 of the thermodynamic temperature of the triple point of water |
| electric current | ampere (A) | The magnitude of the current that, when flowing through each of two long parallel wires of negligible cross-section and separated by 1 m in a vacuum, results in a force between the two wires of $2 \times 10^{-7}$ newton per meter of length |
| luminous intensity | candela (cd) | The luminous intensity in a given direction of a source that emits monochromatic radiation at a frequency of $540 \times 10^{12}$ hertz and that has a radiant intensity in the same direction of 1/683 watt per steradian |
| amount of substance | mole (mol) | The amount of substance that contains as many elementary entities as there are atoms in 0.012 kilogram of carbon-12 |

Table 1.2. Derived quantities and their units in the SI system

| Derived quantity | Name (symbol) | Expression in terms of other SI units | Expression in terms of SI base units |
|---|---|---|---|
| plane angle | radian (rad) | 1 | $m\ m^{-1}$ |
| solid angle | steradian (sr) | 1 | $m^2\ m^{-2}$ |
| frequency | hertz (Hz) | | $s^{-1}$ |
| force | newton (N) | | $m\ kg\ s^{-2}$ |
| pressure, stress | pascal (Pa) | $N/m^2$ | $m^{-1}\ kg\ s^{-2}$ |
| energy, work, quantity of heat | joule (J) | N m | $m^2\ kg\ s^{-2}$ |
| power, radiant flux | watt (W) | J/s | $m^2\ kg\ s^{-3}$ |
| electric charge, quantity of electricity | coulomb (C) | | s A |
| electric potential difference, electromotive force | volt (V) | W/A | $m^2\ kg\ s^{-3}\ A^{-1}$ |
| capacitance | farad (F) | C/V | $m^{-2}\ kg^{-1}\ s^4\ A^2$ |
| electric resistance | ohm ($\Omega$) | V/A | $m^2\ kg\ s^{-3}\ A^{-2}$ |
| electric conductance | siemens (S) | A/V | $m^{-2}\ kg^{-1}\ s^3\ A^2$ |
| magnetic flux | weber (Wb) | V s | $m^2\ kg\ s^{-2}\ A^{-1}$ |
| magnetic flux density | tesla (T) | $Wb/m^2$ | $kg\ s^{-2}\ A^{-1}$ |
| inductance | henry (H) | Wb/A | $m^2\ kg\ s^{-2}\ A^{-2}$ |
| Celsius temperature | degree Celsius (°C) | | K |
| luminous flux | lumen (lm) | cd sr | $m^2\ m^{-2}\ cd = cd$ |
| illuminance | lux (lx) | $lm/m^2$ | $m^2\ m^{-4}\ cd = m^{-2}\ cd$ |
| activity (of a radionuclide) | becquerel (Bq) | | $s^{-1}$ |
| absorbed dose specific energy (imparted) | gray (Gy) | J/kg | $m^2\ s^{-2}$ |
| dose equivalent | sievert (Sv) | J/kg | $m^2\ s^{-2}$ |
| catalytic activity | katal (kat) | | $s^{-1}\ mol$ |

(CIPM), which meets annually and oversees the work of the International Bureau of Weights and Measures (BIPM). The BIPM, based at Sevres just outside Paris, has the responsibility for international standards and is a center for international research and cooperation in metrology.[4] The CIPM has created a number of advisory specialist committees (consultative committees) that are each chaired by a member of CIPM. The committee of interest to chemists is the Consultative Committee on the Amount of Substance (CCQM). It oversees the Avogadro project and coordinates a series of international interlaboratory trials called Key Comparisons (BIPM 2003), which are detailed in chapter 5. The hierarchy of organizations responsible for the Treaty of the Metre is shown in figure 1.4.

### 1.3.3   National Measurement Institutes

Many countries have established national measurement institutes to oversee their metrology systems and obligations to the Treaty of the Metre. Sometimes chemistry and other sciences are separated (as in the National Physical Laboratory and the LGC in the United Kingdom), but increasingly chemical mea-

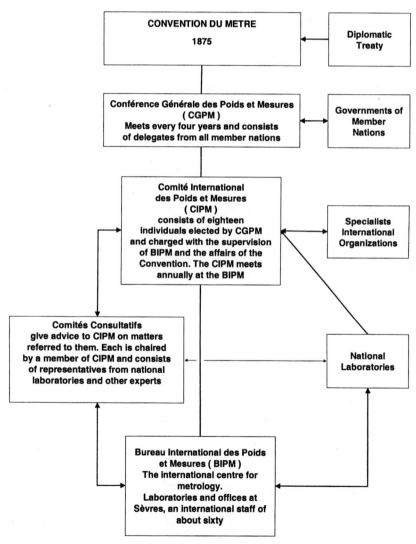

**Figure 1.4.** The Treaty of the Metre and its associated organizations.

surements are seen as an integral part of national metrology. Many of these national measurement institutes take part in the Key Comparisons program of the CCQM (BIPM 2003). An important role of the chemical laboratories in national measurement institutes is to demonstrate a primary measurement capability (i.e. measurements made without reference to other standards) in fields that are deemed important to the country. For example, in sports-mad Australia, the National Analytical Reference Laboratory (NARL) has become a world leader in making reference materials for sports-drugs testing. By demonstrating that these laboratories can make traceable (and therefore comparable) measurements in a wide range of important fields (e.g., trade, health, forensics), a platform for international metrology in chemistry has been set. International cooperation has been enhanced by the formation of regional groupings of national measurement institutes. A list of some national measurement institutes are given in table 1.3, and regional organizations are listed in table 1.4.

### 1.3.4   The SI and Metrological Traceability

Chapter 7 is devoted to metrological traceability, but for now it is important to stress the crucial role that the SI plays in traceability of measurement results. By using the unit mole, kilogram, or meter, there is an implication that the result is indeed traceable to the SI base unit. Although sometimes this might be no more than a forlorn hope, the system of calibrations using certified reference materials is designed to establish the traceability chain and its associated measurement uncertainty. Without an internationally agreed-upon anchor (the SI), measurements made around the world in local systems of units would not be comparable, even if they happened to have the same name or symbol. It is the existence of the SI and its attendant organizational structure that means that 1 kg of rice in China is equivalent to 1 kg of rice in Germany.

## 1.4  Quality Systems

There are many approaches to ensuring quality in an organization. The approach used will depend on the size and nature of the business, the cost of the product, the cost of failure, and the current fashion. It is not my intent to advocate one or another approach. The work of a laboratory is often dictated by accreditation requirements or the law and I offer here what a laboratory can do, not a philosophical framework for quality assurance. Nonetheless, it may be useful to know some of the acronyms and popular trends, so a discerning manager can make a decision about what is best for her or his organization. The following headings do not imply that the methods are mutually exclusive. For example, accreditation might be sought in the context of a TQM approach to quality.

Table 1.3. Some National Metrology Institutes

| Country | Name of the NMI | URL |
|---|---|---|
| Argentina | Instituto Nacional de Tecnología Industrial | http://www.inti.gov.ar/ |
| Australia | National Measurement Institute (NMI) | http://measurement.gov.au/ |
| Austria | Bundesamt für Eich- und Vermessungswesen (BEV) | http://www.bev.gv.at/ |
| Belgium | Service de la Métrologie (SMD) | http://mineco.fgov.be/metrology.en |
| Brazil | The National Institute of Metrology, Standardization and Industrial Quality (InMetro) | http://www.inmetro.gov.br/ |
| Canada | Institute for National Measurement Standards (INMS) | http://inms-ienm.nrc-cnrc.gc.ca |
| Chile | Instituto Nacional de Normalización | http://www.inn.cl/ |
| China | National Institute of Metrology (NIM) | http://en.nim.ac.cn/ |
| Denmark | Danish Safety Technology Authority (SIK) | http://www.danak.dk/ |
| European Commission | Joint Research Centre (JRC) | http://ies.jrc.cec.eu.int/ http://www.irmm.jrc.be/ |
| France | Laboratoire national de métrologie et dessais (LNE) | http://www.lne.fr/ |
| Germany | Federal Institute for Materials Research and Testing (BAM) | http://www.bam.de/ or http://www.ptb.de/ |
| India | National Physical Laboratory | http://www.nplindia.org/ |
| Italy | Consiglio Nazionale delle Ricerche (CNR) | http://www.enea.it/; http://www.ien.it/; or http://www.imgc.cnr.it/ |
| Japan | National Metrology Institute of Japan (NMIJ/AIST) | http://www.aist.go.jp/ |
| Korea | Korea Research Institute of Standards and Science (KRISS) | http://www.kriss.re.kr/ |
| Mexico | Centro Nacional de Metrologia | http://www.cenam.mx/ |
| Netherlands | Netherlands Metrology Service | http://www.nmi.nl/ |

(continued)

Table 1.3. (*continued*)

| Country | Name of the NMI | URL |
|---|---|---|
| New Zealand | Measurement Standards Laboratory, Industrial Research | http://www.irl.cri.nz/msl/ |
| Pakistan | National Physical & Standards Laboratory | http://www.pakistan.gov.pk/divisions |
| Russia | Federal Agency on Technical Regulation and Metrology of Russian Federation (Rostechregulirovanie) | http://www.gost.ru |
| Singapore | Standards, Productivity and Innovation Board (SPRING Singapore) | http://www.psb.gov.sg/ |
| South Africa | CSIR–National Metrology Laboratory | http://www.nml.csir.co.za/ |
| Spain | Spanish Centre of Metrology (CEM) | http://www.cem.es/ |
| Sweden | Swedish National Testing and Research Institute (SP) | http://www.sp.se/ |
| | | http://www.ssi.se/ |
| Switzerland | Swiss Federal Office of Metrology and Accreditation (METAS) | http://www.metas.ch/ |
| United Kingdom | Laboratory of the Government Chemist (LGC) | http://www.lgc.co.uk/ |
| United States | National Institute for Standards and Technology | http://www.nist.gov/ |
| Venezuela | Servicio Autónomo de Normalización, Calidad y Metrología (SENCAMER) | http://www.sencamer.gov.ve/ |

A current comprehensive list of metrology laboratories is given on the NIST web site at http://www.nist.gov/oiaa/national.htm

Table 1.4. Regional groupings of metrology institutes

| Organization | URL |
| --- | --- |
| The Asia Pacific Metrology Programme (APMP) | http://www.apmpweb.org/ |
| Euro-Asian Cooperation of National Metrological Institutions (COOMET) | http://www.coomet.org/ |
| European Collaboration in Measurement Standards (EUROMET) | http://www.euromet.org/ |
| The Inter-American Metrology System (SIM) | http://www.sim-metrologia.org.br/ |
| South African Development Community Cooperation in Measurement Traceability (SADCMET) | http://www.sadcmet.org/ |

### 1.4.1  Accreditation

Chapter 9 is about accreditation to Good Laboratory Practice (GLP; OECD 1998) and ISO/IEC17025 (ISO/IEC 2005), but I discuss here what accreditation actually accredits and its place in the wider scheme of quality systems. There has been much debate, particularly in the pages of the journal *Accreditation and Quality Assurance*, about whether accreditation implies accuracy of measurement results or whether this is an accidental consequence. Certainly the laws of some countries are now specifying that laboratories making particular legal measurements must be accredited to ISO/IEC17025, and accreditation to GLP is a requirement for laboratories around the world that want their results on the stability and toxicity of new drugs to be considered by the U.S. National Institutes of Health. To be accredited to ISO/IEC17025 means that the laboratory's procedures and personnel have been critically reviewed, and that as well as meeting the management quality requirements of the ISO 9000 series, the laboratory methods and practices also meet an ISO standard. Maintenance of accreditation might also require participation in interlaboratory trials (proficiency testing, see chapter 5), which can be a direct demonstration of competence. However, because an inspection that focuses on standard operating procedures and "paper" qualifications of staff can always be open to a certain amount of exaggeration and manipulation, coupled with the accreditation body's desire to give accreditation rather than deny it, accreditation should be seen as a necessary but not sufficient condition for quality. At least in recent years accreditation organizations have an ISO standard (ISO/IEC 2004) to which they, too, must be accredited—"Sed quis custodiet ipsos Custodes," indeed.[5]

### 1.4.2  Peer Review and Visitors

Peer review, initiated by the accreditation body, is the most common method of determining suitability for accreditation. "Peer review" in this case refers to an organization using outsiders as part of its quality system. Popular in academia and government circles, the invitation of selected professionals to investigate an organization has become one means of reviewing overall quality. A review committee is set up with outside participation, and members are asked to address certain issues and questions concerning the organization. Peer review of laboratories is less an ongoing process as it is a one-time procedure directed to a specific issue (accreditation) or problem. If initiated by senior management, peer review is often seen as a prelude to some kind of shake-up in the organization, personnel are typically nervous about the outcome. A properly independent panel can, with a little intelligent probing, can uncover some of the most seemingly intractable problems of a laboratory, and with sensitivity recommend changes that are in the best interests of the institution. The usefulness of peer review rests on the people chosen to make the visit and the willingness and ability of the organization to implement their suggestions. Questions must be addressed such as, do they have the right mix of expertise? Have they been given enough time to complete their tasks? Do they have the authority to see everything they deem relevant? Will the management take any notice of their report?

### 1.4.3  Benchmarking

Benchmarking is like peer review from the inside. In benchmarking an organization goes outside to other organizations that are recognized as leaders in the sector and compares what is being done in the best-practice laboratory with their own efforts. First, the field must have developed enough to have agreed-upon best practices. At universities, for example, the sheer number of chemistry departments means that an institution can readily be found that can act as an exemplar for the procedures. The exemplar may not necessarily be the department that is academically the best. Having many Nobel Laureates that are funded by generous alumni donations may be the dream of every university, but it may be better to benchmark against a midstream university that is of about the same size and funding base. Another consideration is whether the targeted institution will agree to share the secrets of its success, especially if it is seen as a rival for clients, funds, or students. Finally, benchmarking can be very selective. No organizations are the same, and what works well for one might not be the best for another. Management needs to be realistic in the choice of what facets of a benchmarking exercise can be usefully applied in their own business.

### 1.4.4  Total Quality Management

Total quality management became popular in the 1990s, after decades of a move toward taking quality control from the factory floor operator to a management-centered operation (total quality control of Ishikawa in 1950–1960 [Ishikawa 1985, page 215]) to the holistic approach enshrined in standards such as BS 7850 (BS 1992, 1994) and defined in ISO 8402 (ISO 1995) and subsequently ISO 9000 (ISO 2005). Although its lessons have survived and prospered in some areas, the challenge for analytical laboratories is to decide whether this approach is worth the effort, and, if so, what is the best implementation. The basis of the TQM approach involves concepts such as the driver of customer satisfaction, a correct policy and strategy in the organization, accountability of individuals, and continuous improvement. The ISO standard 8402 defines quality in TQM as the achievement of all managerial objectives.

TQM has embraced all the methods of monitoring and testing that are discussed in this book and puts them together in a system that operates enterprise wide. The first principle is an absolute commitment by the highest management that the organization will actively accept the philosophy of TQM and commit resources to its implementation. Each member of the organization is involved in rounds of continuous improvement, with a structured program in which the product and processes are studied, improvements are planned, changes are implemented, and so on. The traditional TQM concerns manufacturing industry where a goal is "zero defects." How this might translate in an analytical laboratory is worthy of discussion. For a pathology laboratory, perhaps "no dead patients" (due to our errors), or for an environmental monitoring laboratory of a chemical plant, "no prosecutions by the EPA." To achieve these lofty aims, there has to be a good understanding of the procedures and practices in the laboratory. This involves validation of methods, measurement uncertainty and traceability, and perhaps monitoring by participation in interlaboratory trials and the use of control charts. One facet of TQM that is very applicable to chemical laboratories is the involvement of personnel at all levels. Usually the technical staff who physically prepare the samples and conduct the analyses are the best people to suggest improvements. They are also the ones who might be cutting corners and creating problems further along the chain. By involving everyone in the rounds of quality improvement, a perspective on what actions are genuinely beneficial can be attained in an atmosphere in which all staff have a positive input. In this respect TQM can be more productive than peer review or benchmarking, in which the personnel can feel they are being judged and so might hide improper practices.

### 1.4.5  Project Management and Six Sigma

Project management and six sigma are approaches based on highly structured planning with monitoring at each stage (iSixSigma 2006). The four headings of project management are defining, planning, implementing, and completing. Project management relies on tools such as Gantt charts and emphasizes effective communication within an organization to achieve the set goals. Six sigma is named after a tolerance spread of $\pm 6\sigma$ which, after allowance for $\pm 1.5\sigma$ for random effects, leads to a calculation of 3.4 defects per million. The tolerance ($T$) is set by the requirements of the customers, and so it then behooves the analyst to achieve a target measurement uncertainty that complies with $T = 6\sigma$. With this target an appropriate testing and action regime can then be put in place. The basic sequence of events associated with six sigma is recognize–define–measure–analyze–improve–control. These events are discussed in two scenarios, one in which an existing product is found in need of improvement after customer concerns, and the other for the development of new products. The first scenario makes use of the acronym DMAIC, which stands for define–measure–analyze–improve–control, and the second scenario uses DMADV, or define–measure–analyze–design–verify. It should be noted that the use of six sigma in the analytical laboratory does have some problems. In a process control laboratory I am familiar with, the laboratory manager was arguing with his six-sigma masters when the software flagged out-of-control data—data that he believed quite reasonable. It is important to understand what sigma is being referred to and to understand whether or not the system is in genuine statistical control. Is sigma the measurement repeatability, the measurement uncertainty, or the overall standard deviation of the product and measurement? Having identified an acceptable tolerance, is mapping ± 6$\sigma$ on to it a feasible proposition, and are the data appropriately normal out to these far-flung regions? I will look at some of these issues in the chapters of this book, but for now treat all quality approaches with some caution until you are satisfied that they do hold some use for your laboratory (and not just the airline, car plant, or electronic manufacturer that are always used as shining examples).

### 1.4.6  The Hibbert Method

There is no complete specification for an analytical result as there is for a toaster. Quality must be inferred from a number of indicators, and a good laboratory manager will understand this and take from popular approaches methods sufficient to yield quality results. The purpose of this book is to help the manager do this, not to prescribe 10 steps to enlightenment. This is not another quality guru speaking, although if you want to read some of these gurus, try their own words rather than those of their disciples (e.g., Ishikawa 1985, Juran 1992). In the words of Brian (eponymous hero of the Monty Python film *Life of Brian*), "Make your own minds up!"

### Notes

1. At the time of completing this book, the third edition of the VIM was in its last draft before publication. Unfortunately it was not released and so cannot be quoted as such. Some definitions in this book lean heavily on the thoughts of the VIM revisers but without attribution. Otherwise the second edition is quoted.

2. Chemists are not alone. The *Mars Climate Orbiter* crashed into Mars in September 1999 following a confusion in programming of software that modeled forces in English units instead of the metric units that were expected by the thruster controllers and other programs. This incident is related in the delightfully titled NASA document "Mars Climate Orbiter Mishap Investigation Report." (Stephenson 1999) The "mishap" cost around $125,000,000.

3. The bodies of the Treaty of the Metre, perhaps in deference to the leading role played by France, are known by their French initials. CGPM = Conférence Général des Poids et Mesures.

4. There is only one artifact remaining as an international standard—the international prototype kilogram. This is a platinum/iridium alloy housed in two bell jars in a vault at the BIPM. By decree and treaty it weighs exactly 1 kg when removed and cleaned according to the specifications of the definition. When the Avagadro constant ($N_A$) is measured or agreed to sufficient precision, it will be possible to redefine the kilogram in terms of the mass of $N_A$ carbon-12 atoms (see chapter 7).

5. Often misquoted, the quote refers to locking up and guarding wives to avoid unfaithfulness [Juvenal, Satires no. 6 1.347].

## References

AOAC International (2006), Terms and Definitions. Available: http://www.aoac.org/terms.htm, accessed 23 September 2006.

BIPM (2003), The BIPM key comparison database (KCDB). Available: www.bipm.org/kcdb.

BIPM (2005), The international system of units. Available: http://www.bipm.fr/en/si/.

BS (1992), Total quality management, part 1: Guide to management principles, 7850-1 (London: British Standards).

BS (1994), Total quality management, part 2: Guide to quality improvement methods, 7850-2 (London: British Standards).

De Bievre, P (1993), How do we prepare for environmental measurements beyond Europe "1993." *International Journal of Environmental Analytical Chemistry,* 52, 1–15.

Deming, W E (1982), *Out of the crisis* (Cambridge, MA: Massachusetts Institute of Technology, Center for Advanced Engineering Study).

Dybkaer, R. (2004), An ontology on property for physical, chemical, and biological systems. *APMIS,* 112 Blackwell Munksgaard 2004 (University of Copenhagen).

Garfield, F M, Klesta, E, and Hirsch, J (2000), *Quality assurance principles for analytical laboratories,* 3rd ed. (Arlington, VA: AOAC International).

Ishikawa, K (1985), *What is total quality control? The Japanese Way,* trans. D J Lu (Englewood Cliffs, NJ: Prentice-Hall).

iSixSigma (2006), iSixSigma web site. http://www.isixsigma.com/ [accessed 22 January 2006].

ISO (1993), *International vocabulary of basic and general terms in metrology,* 2nd ed. (Geneva: International Organization for Standardization).

ISO (1994), Quality management and quality assurance—Vocabulary, 8402 (Geneva: International Organization for Standardization).

ISO (2005), Quality management systems—Fundamentals and vocabulary, 9000 (Geneva: International Organization for Standardization).

ISO/IEC (2004), Conformity assessment—General requirements for accreditation bodies accrediting conformity assessment bodies, 17011 (Geneva: International Organization for Standardization).

ISO/IEC (2005), General requirements for the competence of calibration and testing laboratories, 17025 2nd ed. (Geneva: International Organization for Standardization).

Juran, J M (1992), *Juran on quality by design: the new steps for planning quality into goods and services* (New York: Free Press).

King, B (1995), Quality in the analytical laboratory, in M Parkany (ed.), *Quality assurance and TQM for analytical laboratories* (Cambridge: The Royal Society of Chemistry), 8–18.

LGC (2005), Valid Analytical Measurement Programme web site. http://www.vam.org.uk/. [accessed 21 July 2005]

Massart, D L, Vandeginste, B G M, Buydens, J M C, de Jong, S, Lewi, P J, Smeyers-Verberke, J (1997), *Handbook of chemometrics and qualimetrics part A*, 1st ed., vol 20A in *Data handling in science and technology* (Amsterdam: Elsevier).

OECD, Environment Directorate—Chemicals Group and Management Committee (1998), OECD series on principles of good laboratory practice and compliance monitoring, number 1: OECD Principles on Good Laboratory Practice (Paris: Organisation for Economic Co-operation and Development).

Parkany, M (ed.), (1993), *Quality assurance for analytical laboratories* (Cambridge: Royal Society of Chemistry).

Parkany, M (ed.), (1995), *Quality assurance and TQM for analytical laboratories* (Cambridge: Royal Society of Chemistry).

Prichard, E (1995), *Quality in the analytical chemistry laboratory* (Chichester, UK: John Wiley and Sons).

Roy, Ranjit K (2001), *Design of Experiments Using The Taguchi Approach: 16 Steps to Product and Process Improvement* (New York: John Wiley & Sons) 531.

Sargent, M and MacKay, G (1995), *Guidelines for achieving quality in trace analysis* (Cambridge: Royal Society of Chemistry).

Shewhart, W (1931), *The Economic Control of Quality of Manufactured Products* (New York: D. van Nostrand).

Stephenson, Arthur G (1999), 'Mishap Investigation Board, Phase I Report', (Washington DC: NASA), pp 44.

U.S. Senate (2001), Testimony of Dr. Willie E. May, chief, Analytical Chemistry Division, Chemical Science and Technology Laboratory, NIST, before the Committee on Commerce, Science, and Transportation, Subcommittee on Science, Technology and Space, United States Senate on "E-Health and Technology: Empowering Consumers in a Simpler, More Cost Effective Health Care System" 23 July, 2001.

Wenclawiak, B W, Koch, M, and Hadjicostas, E (eds.) (2004), *Quality assurance in analytical chemistry: Training and teaching* (Heidelberg: Springer Verlag).

# 2

## Statistics for the Quality Control Laboratory

### 2.1 Lies, Damned Lies, and Statistics

Because volumes are devoted to the statistics of data analysis in the analytical laboratory (indeed, I recently authored one [Hibbert and Gooding 2005]), I will not rehearse the entire subject here. Instead, I present in this chapter a brief overview of the statistics I consider important to a quality manager. It is unlikely that someone who has never been exposed to the concepts of statistics will find themselves in a position of QA manager with only this book as a guide; if that is your situation, I am sorry.

Here I review the basics of the normal distribution and how replicated measurements lead to statements about precision, which are so important for measurement uncertainty. Hypothesis and significance testing are described, allowing testing of hypotheses such as "there is no significant bias" in a measurement. The workhorse analysis of variance (ANOVA), which is the foundational statistical method for elucidating the effects of factors on experiments, is also described. Finally, you will discover the statistics of linear calibration, giving you tools other than the correlation coefficient to assess a straight line (or other linear) graph. The material in this chapter underpins the concept of a system being in "statistical control," which is discussed in chapter 4. Extensive coverage of statistics is given in Massart et al.'s (1997) two-volume handbook. Mullins' (2003) text is devoted to the statistics of quality assurance.

## 2.2 Uncertainty in Chemical Measurements

Berzelius (1779–1848) was remarkably farsighted when he wrote about measurement in chemistry: "not to obtain results that are absolutely exact—which I consider only to be obtained by accident—but to approach as near accuracy as chemical analysis can go." Did Berzelius have in mind a "true" value? Perhaps not, and in this he was being very up to date. The concept of a true value is somewhat *infra dig*, and is now consigned to late-night philosophical discussions. The modern approach to measurement, articulated in the latest, but yet-to-be-published *International Vocabulary of Basic and General Terms in Metrology* (VIM; Joint Committee for Guides in Metrology 2007), considers measurement as a number of actions that improve knowledge about the unknown value of a measurand (the quantity being measured). That knowledge of the value includes an estimate of the uncertainty of the measurement result. The assessment of measurement uncertainty is facilitated by taking a view of the nature of each uncertainty component. There are three kinds of errors that contribute to the variability of an analytical result: normally distributed random errors, systematic errors (bias and recovery), and gross errors. Gross errors include spilling a portion of the analyte, using the wrong reagent , and misplacing a decimal point. Results arising from gross errors cannot be considered among measurement results that are the subject of the analysis, but must be duly identified, documented and set aside. Gross errors will not be discussed further. Random errors can be treated by statistics of the normal distribution, and systematic errors can be measured and corrected for.

### 2.2.1  Systematic versus Random Errors

The measurement uncertainty, quoted as a confidence interval about the measured value, is a statement of the extent of the knowledge concerning the value of the measurand. Nevertheless, calculating the uncertainty range is influenced by the nature of the effects contributing to uncertainty: one must consider whether an uncertainty contribution can be described by a normal distribution (i.e., deemed to be random), or whether it is considered a systematic error. Many factors that affect uncertainty are just about random: temperature fluctuations, an analyst's shaky hand while filling a pipette to the mark, or voltage noise affecting the rate of arrival of ions in a mass spectrometry source are some examples. The reason that the modern approach, to some extent, blurs the distinction between random and systematic error is that it is easy to turn one form of error into the other. For example, a 10.00-mL pipette might be sold with a tolerance of ±0.02 mL. This means that the manufacturer assures users that the pipette will never deliver < 9.98 mL and never > 10.02 mL. Suppose the volume of a particular pipette is 9.99 mL. If an experiment were repeated a number of times using this pipette, all the results would be affected equally by the 0.01-mL short delivery. Unless the pipette were independently calibrated and the more correct volume of 9.99 mL used in the calculations, then the systematic error in the results would be there forever. In contrast, if for each repeat of the measurement a different pipette were used, then it might be expected that after enough repeats with one pipette delivering 10.015 mL, and another 9.987 mL, and so on, the mean delivery should tend toward 10.00 mL, and the standard deviation of the results would be greater than if a single pipette were used. The systematic error of using a single, uncalibrated pipette, which gives a bias in the result, has been turned, by using many pipettes, into a random error with less bias, but greater variance.

### 2.2.2 Repeatability and Reproducibility

"Precision" is the word used to describe the spread of a number of results, and it is usually expressed as a standard deviation ($\sigma$) or relative standard deviation (RSD; see below). "Good precision" means that the results tend to cluster together with an appropriately small $\sigma$, "poor precision" implies that the data are spread widely and that $\sigma$ is large compared with the measurement result. In chemistry, it is important to make clear under what conditions the results have been replicated when quoting an estimate of precision. The more conditions have been allowed to vary, the greater the spread of results. Repeatability conditions allow the minimum amount of variation in the replicated experiments. The same analyst performs experiments on the same equipment, using the same batch of reagents, within a short time. Repeatability ($\sigma_r$) is the closeness of agreement between the results of successive measurements of the same measurand carried out subject to the following conditions:

- the same measurement procedure
- the same observer
- the same measuring instrument used under the same conditions
- the same location
- repetition over a short period of time (ISO 1994, terms 3.13 and 3.14)

Repeatability is expressed as a standard deviation (described below).

As soon as more of the conditions are changed, then the standard deviation of results is known as the reproducibility. Reproducibility ($\sigma_R$) is the closeness of agreement between the results of measurements of the same measurand, where the measurements are carried out under changed conditions, such as:

- principle or method of measurement
- observer
- measuring instrument
- location
- conditions of use
- time (ISO 1994, terms 3.17 and 3.18)

Any or all of these conditions can be varied. To provide some guidance, "intralaboratory reproducibility" is used to express changes only within a laboratory, and "interlaboratory reproducibility" is used to refer to the changes that occur between laboratories, for example in proficiency testing, interlaboratory method validation studies, and the like. Interlaboratory reproducibility is usually two to three times the repeatability.

## 2.3 Basics: Describing the Normal Distribution

### 2.3.1 The Normal Distribution

The normal, or Gaussian, distribution occupies a central place in statistics and measurement. Its familiar bell-shaped curve (the probability density function or pdf, figure 2.1) allows one to calculate the probability of finding a result in a particular range. The x-axis is the value of the variable under consideration, and the y-axis is the value of the pdf.

$$f(x) = \frac{1}{\sigma\sqrt{2\pi}} \exp\left[-\frac{(x-\mu)^2}{2\sigma^2}\right]$$  (2.1)

An area under the pdf is the probability of a result between the $x$ values defining the area. Figure 2.2 shows the area from $-\infty$ to $x$.

$$F(x_1 > x > x_2) = \int_{x=x_1}^{x=x_2} \frac{1}{\sigma\sqrt{2\pi}} \exp\left[-\frac{(x-\mu)^2}{2\sigma^2}\right] dx$$  (2.2)

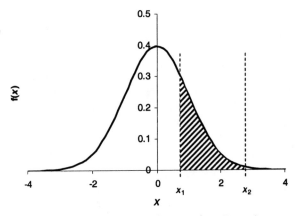

**Figure 2.1.** The standardized normal or Gaussian distribution. The shaded area as a fraction as the entire area under the curve is the probability of a result between $x_1$ and $x_2$.

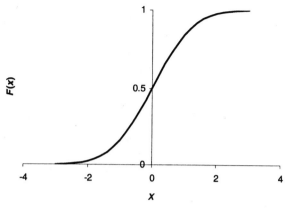

**Figure 2.2.** The cumulative normal distribution.
$F(x)$ is the probability of finding a result
between $-\infty$ and $x$.

The normal distribution is characterized by two parameters. The center
of the distribution, the mean, is given the symbol $\mu$. The width of the distribution is governed by the variance, $\sigma^2$. The square root of the variance, $\sigma$, is
the standard deviation.

Why is any of this of interest? If it is known that some data are normally
distributed and one can estimate $\mu$ and $\sigma$, then it is possible to state, for
example, the probability of finding any particular result (value and uncertainty range); the probability that future measurements on the same system
would give results above a certain value; and whether the precision of the
measurement is fit for purpose. Data are normally distributed if the only
effects that cause variation in the result are random. Random processes are
so ubiquitous that they can never be eliminated. However, an analyst might
aspire to reducing the standard deviation to a minimum, and by knowing
the mean and standard deviation predict their effects on the results.

In any experiment, therefore, all significant systematic errors should be
measured and corrected for, and the random errors, including those pertain-
ing to the bias corrections, estimated and combined in the measurement
uncertainty.

Later, how to calculate the probability of finding results that are removed
from the mean will be explained. This is the basis of hypothesis testing. Once
the form of the distribution is known, a desired probability is then an area
under the pdf. Testing at 95% probability usually entails determining the
limits that encompass 95% of the distribution in question. There are two
ways of doing this: symmetrically about the mean, or starting at infinity in
one direction or the other. Figure 2.3 shows the difference. Most tests use
symmetrical boundaries; there is usually no reason to expect an outlier, for
example, to be much greater or much less than the mean. In cases where

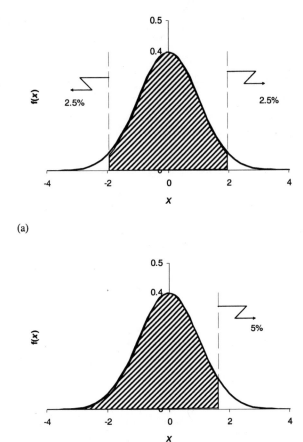

(a)

(b)

**Figure 2.3.** Ninety-five percent of a standardized normal distribution. (a) Two tailed: 2.5% lies at $z < -1.96$, and $z > +1.96$. (b) One tailed: 5% lies at $z > 1.65$.

only one side of the distribution is important (e.g., testing against an upper limit of a pollutant), the test is done with all the probability at the end in question (figure 2.3b).

### 2.3.2   Mean and Standard Deviation

The normal distribution of figure 2.1 is the theoretical shape of a histogram of a large number of estimations. In chemistry often only duplicate measurements are done, or in full method-validation mode, a dozen or so. Herein lies the problem. The chemist does not have enough information to report the values of the mean and standard deviation of the *population* of experiments from which the results are drawn. Instead, we calculate the sample mean ($\bar{x}$) and sample standard deviation ($s$)

$$\bar{x} = \frac{\sum\limits_{i=1}^{i=n} x_i}{n} \tag{2.3}$$

$$s = \sqrt{\frac{\sum\limits_{i=1}^{i=n}(x_i - \bar{x})^2}{n-1}} \tag{2.4}$$

In Microsoft Excel, these are calculated by the functions =AVERAGE (*range*) and =STDEV (*range*), respectively, where range is the range of cells containing the $n$ data (e.g., A1:A5). It is said that $\bar{x}$ is an estimate of $\mu$ and $s$ is an estimate of $\sigma$.

In analytical chemistry, the ratio of the standard deviation to the mean, or RSD is often quoted. This quantity is also known as the coefficient of variation (CV), but the term is no longer favored and RSD should be used. Remember both the mean and the standard deviation have the units of the measurand. The relative standard deviation has unit one and is usually expressed as a percentage.

### 2.3.2.1  Standard Deviation of the Mean

The mean and standard deviation of $n$ results is given in equations 2.3 and 2.4, respectively. Suppose now that this set of $n$ experiments is repeated a large number of times. The average of all the values of $\bar{x}$ tends to the population mean, $\mu$. The standard deviation of the values of $\bar{x}$ is $\sigma/\sqrt{n}$. This is a very important result, and the difference between the standard deviation and the standard deviation of the mean must be understood. The standard deviation calculated from all the $n$ individual data values would tend to the standard deviation of the population ($\sigma$), but as $n$ becomes greater the standard deviation of the means would become smaller and smaller and tend to zero, by virtue of $\sigma$ being divided by $\sqrt{n}$. As expected, the more replicates that are averaged, the greater the confidence that the calculated sample mean is near the mean of the population. Even when only a few results are averaged, a *sample* standard deviation of the mean may be calculated as $s/\sqrt{n}$. As I show in the next section, when the sample standard deviation of the mean is used to calculate confidence intervals, some allowance will have to be made for the fact that $s$ is only an estimate of $\sigma$. Another feature of averages is that the means of $n$ measurements tend to a normal distribution as $n$ becomes greater, even if the population is not normally distributed. This possibly surprising result is very useful for measurement scientists, as it allows them to assume normality with a certain amount of abandon, if a suitable number of replicates are used to provide a mean result.

### 2.3.2.2  Standardized Normal Variable: z Score

When discussing normally distributed variables, a convenient way of expressing the data is the distance from the mean in standard deviations. This is known as the standardized normal variable or z value. A result that has $z = -1.2$ is 1.2 times $\sigma$ less than the mean.

$$z_i = \frac{(x_i - \mu)}{\sigma} \qquad (2.5)$$

For data that are to be compared with expected limits of a distribution (e.g., in an interlaboratory trial), the z score is calculated using either the mean and sample standard deviation of the data,

$$z_i = \frac{(x_i - \bar{x})}{s} \qquad (2.6)$$

or the assigned reference value of the measurand, $x_a$ , and target standard deviation $(\sigma_{target})$,

$$z_i = \frac{(x_i - x_a)}{\sigma_{target}} \qquad (2.7)$$

or one of the robust estimates described below.

### 2.3.2.3  Robust Estimates

If data are normally distributed, the mean and standard deviation are the best description possible of the data. Modern analytical chemistry is often automated to the extent that data are not individually scrutinized, and parameters of the data are simply calculated with a hope that the assumption of normality is valid. Unfortunately, the odd bad apple, or outlier, can spoil the calculations. Data, even without errors, may be more or less normal but with more extreme values than would be expected. These are known has heavy-tailed distributions, and the values at the extremes are called outliers. In interlaboratory studies designed to assess proficiency, the data often have outliers, which cannot be rejected out of hand. It would be a misrepresentation for a proficiency testing body to announce that all its laboratories give results within ± 2 standard deviations (except the ones that were excluded from the calculations).

   Consider these data for an acid base titration: 10.15 mL, 10.11 mL, 10.19 mL, 11.02 mL, 10.11 mL (see figure 2.4). The result 11.02 mL is almost certainly a typo and should be 10.02 mL, but suppose it goes unnoticed. The mean

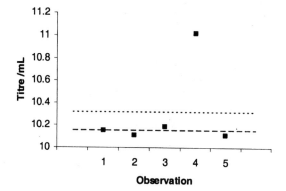

(a)

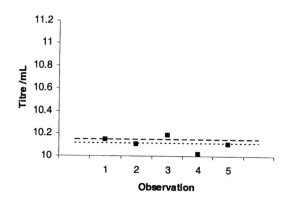

(b)

**Figure 2.4.** (a) Set of titration volumes with outlier. Dotted line is the mean. Dashed line is the median. (b) Set of titration volumes with outlier corrected from 11.02 mL to 10.02 mL on the same scale.

is dragged up to 10.32 mL, and the sample standard deviation is a whopping 0.39 mL. Subsequent calculations would not represent the true picture of the concentration of acid. One method of allowing for outliers when calculating estimates of the mean and standard deviation is to use a robust estimate of the mean (the median) and a robust estimate of the standard deviation (the median absolute deviation divided by 0.6745). The median of a set of data is the middle value when the results are sorted by order of magnitude. If there is an odd number of data, the median is one of the values (e.g., out of five data, the value of the third is the median: 1, 2, **3**, 4, 5). If there is an even number of data, the median is the average of the two middle values (e.g., with six values, the third and fourth are averaged: 1, 2, [**3** + **4**]/2, 5, 6). For normally distributed data, because the distribution is symmetrical, the mean, $\mu$, and the median are the same. For small data sets with a heavy tail it has been shown that the mean is skewed in the direction of the outlier. The median is not so affected, as the end value is the end value, whatever its magnitude.

The median of the titrations in the example above is 10.15 mL, which is a much more representative value. If the result 11.02 mL is replaced by 10.02 mL (this would require evidence that there had been an error in transcribing results), then the mean becomes the reasonable 10.12 mL and the standard deviation is 0.06 mL.

The calculation of the mean absolute deviation (MAD) is

$$MAD_i = \text{median} \left( | x_i - \text{median}(x_i) | \right) \qquad (2.8)$$

As may be seen in the example in spreadsheet 2.1, the robust estimate of standard deviation (MAD/0.6745) agrees with standard deviation of the amended data.

In interlaboratory trials that are not informed by an assigned value and target standard deviation, a robust $z$ score is used to account for all the available data

$$z_i = \frac{x_i - \text{median}(x_i)}{\sigma_{\text{robust}}} \qquad (2.9)$$

| F17 | | $f_x$ | | | |
|---|---|---|---|---|---|
| | A | B | C | D | E |
| 6 | | Titre /mL $x_i$ | median($x_i$ - median($x$)) | Titre /mL $x_i$ | median($x_i$ - median($x$)) |
| 7 | 1 | 10.15 | 0 | 10.15 | 0.04 |
| 8 | 2 | 10.11 | 0.04 | 10.11 | 0 |
| 9 | 3 | 10.19 | 0.04 | 10.19 | 0.08 |
| 10 | 4 | 11.02 | 0.87 | 10.02 | 0.09 |
| 11 | 5 | 10.11 | 0.04 | 10.11 | 0 |
| 12 | | | | | |
| 13 | Mean /mL | 10.32 | | 10.12 | |
| 14 | s /mL | 0.39 | | 0.063 | |
| 15 | median /mL | 10.15 | | 10.11 | |
| 16 | MAD/0.6745 /mL | | 0.059 | | 0.059 |
| 17 | | | | | |

**Spreadsheet 2.1.** Calculation of mean, standard deviation, median, and mean absolute deviation of titration data with outlier, and after correction (shaded cells show the change from 11.02 to 10.02).

The robust estimate of the standard deviation can be the MAD defined above, or if there are sufficient data, 1.35 × interquartile range. The interquartile range (IQR) is the range spanning the middle 50% of the data. However, chemistry rarely has the luxury of sufficient data for a meaningful calculation of the IQR.

In statistical process control when only a small set of data is available, the range of the data is often used to give an estimate of the standard deviation. The range of a set of $n$ data is divided by a parameter, $d_n$ (also called $d_2$).

$$\sigma_{est} = \frac{\left(x_{max} - x_{min}\right)}{d_n}$$

$$\tag{2.10}$$

Table 2.1 gives values up to $n = 10$, although with this number of data it would be advisable to simply compute the sample standard deviation directly. The values of $d_n$ are nearly $\sqrt{n}$, and this is sometimes used in a quick calculation. A better estimate of $d_n$ up to $n = 10$ is

$$d_n = 2.44 \times \sqrt{n} - 0.32 \times n - 1.38 \tag{2.11}$$

### 2.3.3  Confidence Intervals

A sample standard deviation tells something about the dispersion of data, but a statement with a probability may be preferred. A result such as 1.23 ± 0.03 ppm means different things to different people. Even if it is annotated by "(95% confidence interval)," there is still some doubt about its interpretation. The use of plus or minus should refer to a high probability (95% or 99%) range; it should not be used to state a standard deviation. If the standard deviation is to be given, then write "the mean result was 1.23 ppm ($n = 5$, $s = 0.04$ ppm)," and everyone will understand what is meant. If an expanded uncertainty from an estimated measurement uncertainty (see chapter 6) appears after the plus or minus sign, then there should be an appropriate accompanying statement. For example, "Result: 1.23 ± 0.03 ppm, where the reported uncertainty is an expanded uncertainty as defined in the VIM, 3rd edition (Joint Committee for Guides in Metrology 2007), calculated with a coverage factor of 2, which gives a level of confidence of

Table 2.1. Values of $d_n$ used in estimating standard deviation ($s$) from the range ($R$) of one set of data, $s \approx R/d_n$

| | | | | $n$ | | | | |
|---|---|---|---|---|---|---|---|---|
| | 2 | 3 | 4 | 5 | 6 | 7 | 8 | 10 |
| $d_n$ | 1.41 | 1.91 | 2.24 | 2.48 | 2.67 | 2.82 | 2.95 | 3.16 |

Adapted from Oakland (1992).

approximately 95%." Some of these words are optional, and expanded uncertainty is not actually defined here, but the thrust of the statement is that the range about the result is where the value of the measurand is thought to lie with the stated confidence.

However, the GUM (*Guide to the Expression of Uncertainty of Measurement*) approach (ISO 1993a), which leads to the verbose statement concerning expanded uncertainty quoted above, might not have been followed, and all the analyst wants to to do is say something about the standard deviation of replicates. The best that can be done is to say what fraction of the confidence intervals of repeated experiments will contain the population mean. The confidence interval in terms of the population parameters is calculated as

$$\mu \pm \frac{z_\alpha \sigma}{\sqrt{n}}$$
(2.12)

where $\pm z_\alpha$ is the range about $\mu$ of the standardized normal distribution encompassing $100(1-\alpha)\%$ of the distribution. For a range with $z = 2$, approximately 95% of the distribution is contained within $\mu \pm 2$ standard deviations. When only a sample mean and standard deviation are available, the confidence interval is

$$\bar{x} \pm \frac{t_{\alpha'',n-1} s}{\sqrt{n}}$$
(2.13)

where $t_{\alpha'',n-1}$ is the two-tailed Student's $t$ value for $100(1-\alpha)\%$ at $n-1$ degrees of freedom.[1] The added complication of using the $t$ value is that the confidence interval, when calculated from the sample standard deviation ($s$) needs to be enlarged to take account of the fact that $s$ is only an estimate of $\sigma$ and itself has uncertainty. The degrees of freedom are necessary because the estimate is poorer for smaller sample sizes. Table 2.2 shows values of $t$ for 95% and 99% confidence intervals.

When $n$ becomes large the $t$ value tends toward the standardized normal value of 1.96 ($z = 1.96$), which was approximated to 2 above. The 95% confidence interval, calculated by equation 2.13, is sometimes explained much like the expanded uncertainty, as a range in which the true value lies with 95% confidence. In fact, the situation is more complicated. The correct statistical statement is: "if the experiment of $n$ measurements were repeated under identical conditions a large number of times, 95% of the 95% confidence intervals would contain the population mean."

Table 2.2 also reveals that if only two or three experiments are performed, the resulting confidence intervals are so large that there may be little point in doing the calculation. Usually there will be some knowledge of the re-

Table 2.2. Values of the two-tailed Student's $t$
distribution calculated in Excel by =TINV(0.05,
n − 1) and =TINV(0.01, n − 1)

| | | Student's $t$ value | |
| --- | --- | --- | --- |
| $n$ | Degrees of freedom $(= n - 1)$ | For 95% confidence $(t_{0.05'',n-1})$ | For 99% confidence $(t_{0.01'',n-1})$ |
| 2 | 1 | 12.71 | 63.66 |
| 3 | 2 | 4.30 | 9.92 |
| 4 | 3 | 3.18 | 5.84 |
| 5 | 4 | 2.78 | 4.60 |
| 6 | 5 | 2.57 | 4.03 |
| 7 | 6 | 2.45 | 3.71 |
| 8 | 7 | 2.36 | 3.50 |
| 9 | 8 | 2.31 | 3.36 |
| 10 | 9 | 2.26 | 3.25 |
| 50 | 49 | 2.01 | 2.68 |
| 100 | 99 | 1.98 | 2.63 |
| $\infty$ | $\infty$ | 1.96 | 2.58 |

peatability of the experiment from method validation studies done at many
more degrees of freedom, and if the duplicate and triplicate experiments are
consistent with that repeatability, it may be used to calculate the confidence
interval, with a $t$ of 2 for the 95% confidence interval.

When the repeatability standard deviation is known, it can be used to
assess results done in duplicate. If the repeatability is $\sigma_r$, then the 95% confidence interval of the difference of two results is

$$r = \sqrt{2} \times 1.96 \times \sigma_r = 2.8 \times \sigma_r \qquad (2.14)$$

This is a useful check of day-to-day duplicated results. If the difference
is greater than the repeatability limit, $r$, it is advisable to perform another
experiment and test further (see ISO 1994).

## 2.4 Testing

### 2.4.1 Hypotheses and Other Beliefs

If the results of analytical chemistry are used as the bases for decisions, a client will frequently compare the measurement result with some other value.
In process control, a company will have release specifications that must be

met by the production samples; a regulatory authority will test product, effluent, and emissions against legal limits; and trading partners will compare analyses made by the buyer and seller of goods. For samples that clearly pass or fail the test, there is no great need to invoke statistics (although "clearly" is only understood in relation to the measurement uncertainty). However, when results are close to the limit, a knowledge of the uncertainty of measurement is essential to correctly interpret compliance or risk. A possible rule is that a mean value plus its 95% confidence interval cannot exceed a certain limit to be considered compliant (positive compliance) or that both the mean and 95% confidence interval have to exceed the limit to be noncompliant (positive noncompliance). The default (or null) hypothesis that is being tested in each case is "the item does not comply" (positive compliance required) or "the item does comply" (positive noncompliance). The null hypothesis (represented as $H_0$) is considered to hold unless there is convincing evidence to the contrary, at which point it is rejected. Figure 2.5 shows different situations with respect to compliance near limits.

As the probability associated with test data and a control limit can be calculated in Excel, rather than give confidence intervals for arbitrary coverage (95%, 99%), I recommend giving the probability that the limit would be exceeded in repeated measurements. Given a knowledge of the distribution describing results (normal, or $t$ distribution), the areas under the pdf either side of the control limit give the probabilities of complying or not complying (see figure 2.6).

### 2.4.2   General Procedures for Testing Data

#### 2.4.2.1   Testing Against a Given Probability

Reports often state that an effect was "significant at the 95% probability level" or use an asterisk to denote significance at 95% and double asterisks for significance at 99%. Significance at the 95% level means that the null hypothesis ($H_0$ = the effect is not significant) has been rejected because the probability that the test statistic, which has been calculated from the data, could have come from a population for which $H_0$ is true has fallen below 5%. Most statistical tests have the same sequence of events:

1. Formulate the null hypothesis, $H_0$, for which an appropriate statistic can be calculated.
2. Calculate test statistic from the data (call it $x_{data}$).
3. Look up the critical value of the statistic at the desired probability and degrees of freedom (call it $x_{crit}$).
4. If $x > x_{crit}$, then reject $H_0$. If $x < x_{crit}$, then accept $H_0$.

Accepting $H_0$ can be often seen more as recognizing the case against $H_0$ as "not proven" than as definite support for it. It is sometimes reported, "there is not enough evidence to reject $H_0$."

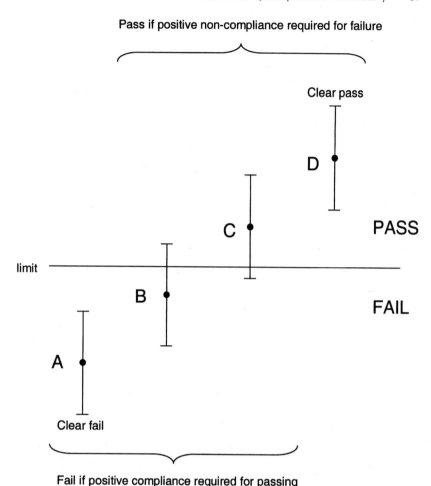

**Figure 2.5.** Measurement results and their measurement uncertainties near a compliance limit. A clearly fails, and D clearly passes. B does not fail if positive noncompliance is required for failure, but otherwise fails. C does not pass if positive compliance is required, but otherwise does pass.

### 2.4.2.2   Calculating the Probability of the Data

For common statistics, such as the Student's $t$ value, chi-square, and Fisher $F$, Excel has functions that return the critical value at a given probability and degrees of freedom (e.g., =TINV(0.05,10) for the two-tailed $t$ value at a probability of 95% and 10 degrees of freedom), or which accept a calculated statistic and give the associated probability (e.g., =TDIST(t, 10, 2) for 10 degrees of freedom and two tails). Table 2.3 gives common statistics calculated in the course of laboratory quality control.

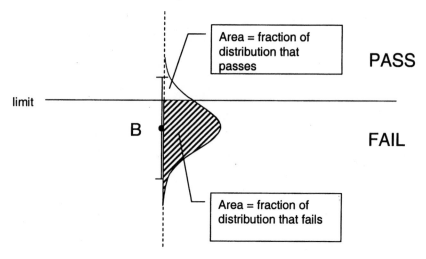

**Figure 2.6.** Distribution about a mean in relation to a control limit. The areas indicated are the fractions of the distributions passing and failing, and so can be equated to the probabilities of the mean result B with 95% confidence interval, complying or not complying.

The advantage of calculating the probability directly, rather than testing against a given probability, is that it is much more informative. A $t$ of 3.99 at 6 degrees of freedom is greater than the critical, two-tailed value at the 95% level ($t_{0.05",6} = 2.45$), and so $H_0$ would be rejected at this level of probability, but discovering that the probability (calculated as =TDIST(3.99,6,2) Pr($|T| > 3.99$)) = 0.0072 tells us that $H_0$ would be rejected at the 99% level and indeed is significant at the 99.3% level.

It must be stressed, because this error is often made ,that the probability calculated above is *not* the probability of the truth of $H_0$. It is the probability that, given the truth of $H_0$, in a repeated measurement an equal or more extreme value of the test statistic will be found.

### 2.4.3   Testing for Normality

I have already pointed out that the distribution of means of the results of repeated measurements tend to the normal distribution when the number of repeats contributing to the mean becomes large. However, the presence of outliers in heavy-tailed distributions when there are not many replicate measurements suggests that from time to time the analyst might need to justify an assumption of normality. When significant effects are being sought using an experimental design (see chapter 3), the underlying assumption is that measured effects are not significant and therefore follow a normal distribution about zero. Results that violate this assumption are deemed significant.

**Table 2.3.** Examples of test statistics used in laboratory testing

| Statistic | Example calculation | $H_0$ | Excel functions |
|---|---|---|---|
| Student's $t$ | $$t = \frac{|\bar{x} - \mu|\sqrt{n}}{s}$$ | The mean $\bar{x}$ comes from a population of means with mean $= \mu$ | $=\text{TINV}(\alpha, n - 1)$<br>$=\text{TDIST}(t, n - 1,$ tails$)$ |
| Chi-square | $$\chi^2 = \frac{(n-1)s^2}{\sigma^2}$$ | The sample variance $s^2$ comes from a population with variance $\sigma^2$ | $=\text{CHIINV}(\alpha, n - 1)$<br>$=\text{CHIDIST}(\chi^2, n - 1)$ |
| Fisher $F$ | $$f = \frac{s_1^2}{s_2^2},$$ where $s_1 > s_2$ | The variances $s_1^2$ and $s_2^2$ come from populations with equal variances | $=\text{FINV}(\alpha, n_1 - 1,$ $n_2 - 1)$<br>$=\text{FDIST}(F, n_1 - 1,$ $n_2 - 1)$ |
| Grubbs' $G$ for single outlier | $$g_i = \frac{|x_i - \bar{x}|}{s}$$ | The extreme data point $x_i$ is not an outlier and comes from a normally distributed population with sample mean $\bar{x}$ and standard deviation $s$ | No Excel functions |
| Grubbs' $G$ for pairs of outliers | $$g(\text{high}) = \frac{\sum_{i=1}^{i=n-2}(x_i - \bar{x}_{1...n-2})^2}{\sum_{i=1}^{i=n}(x_i - \bar{x}_n)^2}$$ $$\text{or } g(\text{high}) = \frac{\sum_{i=3}^{i=n}(x_i - \bar{x}_{3...n})^2}{\sum_{i=1}^{i=n}(x_i - \bar{x}_n)^2}$$ | The extreme pair of data are not outliers and come from a normally distributed population with sample mean $\bar{x}_n$ | No Excel functions |
| Cochran test for homogeneity of variance of $k$ laboratories | $$C = \frac{s_{max}^2}{\sum_{i=1}^{i=k} s_i^2}$$ | The variances of sets of data come from a normal population ($s$ is difference between the results for 2 values) | No Excel functions |

In each case $\bar{x}$ and $s$ are the mean and standard deviation of $n$ data.

There are several tests for normality, most requiring statistical manipulation of the data. If a suitable package is available, you should consider using a chi-square test for large samples ($n > 50$) and the Kolmogorov-Smirnov test for smaller samples. Here I describe a visual test that can easily be implemented in Excel. In the Rankit method a graph is constructed with the ordered results plotted against the expected distribution if they were normal. A straight line indicates normality, with outliers far off the line. The Rankit plot also allows other types of non-normal distribution to be identified. With the data in a column in a spreadsheet, the procedure is as follows for $n$ data:

1. Sort data in increasing order of magnitude.
2. In the next column enter the "cumulative frequency" of each value (i.e., how many data have an equal or lesser value).
3. In the next column calculate the "normalized cumulative frequency," $f$ = cumulative frequency/$(n + 1)$.
4. In the next column calculate the $z$ value =NORMSINV(f).
5. Plot each $z$ against the value.

Spreadsheet 2.2 and figure 2.7 give an example of 10 titration values. Notice how the ranking is done. The Excel function =RANK(x, range, direction) does not quite work, as ties are given the lower number, not the higher. If there are sufficient data to warrant automating this step, the cal-

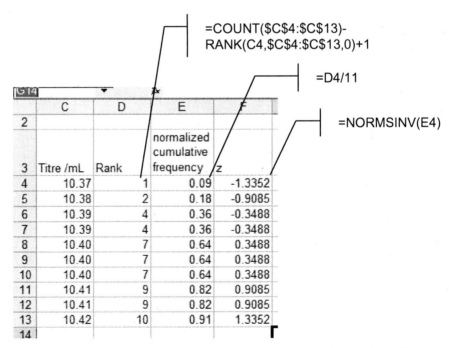

=COUNT($C$4:$C$13)-RANK(C4,$C$4:$C$13,0)+1

=D4/11

=NORMSINV(E4)

| | C | D | E | F |
|---|---|---|---|---|
| 2 | | | | |
| 3 | Titre /mL | Rank | normalized cumulative frequency | z |
| 4 | 10.37 | 1 | 0.09 | -1.3352 |
| 5 | 10.38 | 2 | 0.18 | -0.9085 |
| 6 | 10.39 | 4 | 0.36 | -0.3488 |
| 7 | 10.39 | 4 | 0.36 | -0.3488 |
| 8 | 10.40 | 7 | 0.64 | 0.3488 |
| 9 | 10.40 | 7 | 0.64 | 0.3488 |
| 10 | 10.40 | 7 | 0.64 | 0.3488 |
| 11 | 10.41 | 9 | 0.82 | 0.9085 |
| 12 | 10.41 | 9 | 0.82 | 0.9085 |
| 13 | 10.42 | 10 | 0.91 | 1.3352 |
| 14 | | | | |

**Spreadsheet 2.2.** Example of calculations for a Rankit plot to test normality.

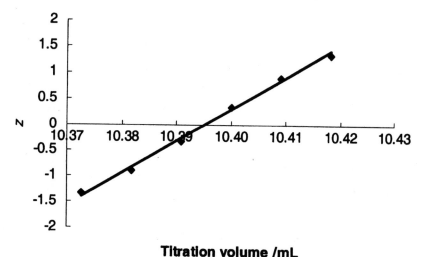

**Titration volume /mL**

**Figure 2.7.** Rankit plot of the data in spreadsheet 2.2.

culation is = n - RANK(x, range, 0) + 1. NORMSINV(f) is the inverse of the standardized normal distribution; that is, it returns the $z$ score of an area under the cumulative normal distribution.

The data fall on a straight line and are concluded to be normally distributed. Outliers in the data are seen as points to the far right and far left of the line.

### 2.4.4  Outliers

If the data as a whole appear normally distributed but there is concern that an extreme point is an outlier, it is not necessary to apply the Rankit procedure. The Grubbs's outlier test (1950) is now recommended for testing single outliers, replacing Dixon's Q-test. After identifying a single outlier, which, of course, must be either the maximum or minimum data value, the $G$ statistic is calculated:

$$g_i = \frac{|x_i - \bar{x}|}{s}$$

(2.15)

There is no easy spreadsheet calculation of the probability associated with $g$, so you must compare it against tables of the critical $G$ value for $n$ data and 95% or 99% probability (table 2.4).

In an example based on IMEP 1 (see chapter 5), lithium in a serum sample was measured six times with the results 0.080, 0.080, 0.100, 0.025, 0.070, and 0.062 mM. Is any result an outlier? A graph of these results (figure 2.8) points to the value 0.025 mM as highly suspicious.

Table 2.4. Values of the critical two-tailed Grubbs's G statistic at 95% and 99% probability for a single outlier

| $n$ | $G_{0.05'',n}$ | $G_{0.01'',n}$ |
|-----|------------|------------|
| 3   | 1.154      | 1.155      |
| 4   | 1.481      | 1.496      |
| 5   | 1.715      | 1.764      |
| 6   | 1.887      | 1.973      |
| 7   | 2.020      | 2.139      |
| 8   | 2.127      | 2.274      |
| 9   | 2.215      | 2.387      |
| 10  | 2.290      | 2.482      |
| 25  | 2.822      | 3.135      |
| 50  | 3.128      | 3.482      |
| 100 | 2.800      | 3.754      |

The mean of these values is 0.0695 mM and the standard deviation is 0.0252 mM, which gives

$$g = \frac{|0.025 - 0.0695|}{0.0252} = 1.762 \tag{2.16}$$

Table 2.4 gives $G_{crit}$ = 1.887. As $g < G_{crit}$, the null hypothesis cannot be rejected and so the point is not an outlier at the 95% probability level. However, in this case $g$ is close to $G_{crit}$, and there might be legitimate concern about this suspicious value.

This Grubbs's test is appropriate only for single outliers. Grubbs also published a test for a pair of outliers at either end of the data. The test statistic is the ratio of the sum of squared deviations from the mean for the set minus the pair of suspected outliers and the sum of squared deviations from the mean for the whole set:

$$g(\text{high pair}) = \frac{\sum_{i=1}^{i=n-2}\left(x_i - \bar{x}_{1...n-2}\right)^2}{\sum_{i=1}^{i=n}\left(x_i - \bar{x}_{1..n}\right)^2}$$

$$\text{or } g(\text{low pair}) = \frac{\sum_{i=3}^{i=n}\left(x_i - \bar{x}_{3...n}\right)^2}{\sum_{i=1}^{i=n}\left(x_i - \bar{x}_{1..n}\right)^2} \tag{2.17}$$

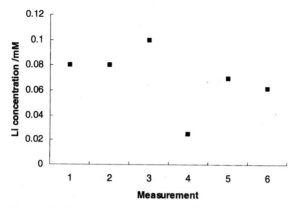

**Figure 2.8.** Data from replicate analyses of Li in serum. Point 4 is a candidate outlier.

Note that the mean used in each calculation is the mean of the set being considered, either all the data in the denominators, or less the two greatest or two least. The paired $G$ statistic is compared with critical values (see table 2.5), and, unusually, the null hypothesis is rejected if the calculated $g$ is *smaller* than the tabulated $G$.

If there are more suspect outliers, perhaps some scrutiny should be directed at the data set as a whole and the data tested for normality.

### 2.4.5  Testing Variances

#### 2.4.5.1  F test for Equality of Variances

Given two sets of repeated measurements, the question of whether the data come from populations having equal variances might arise. This is tested by calculating the Fisher $F$ statistic, which is defined as

$$F = \frac{s_1^2}{s_2^2} \tag{2.18}$$

where the greater value is chosen as $s_1$. The test is then whether $F$ is significantly greater than 1 (which is the expected value if $\sigma_1 = \sigma_2$). In Excel the critical values of the one-tailed $F$ distribution at probability $\alpha$ are given by =FINV($\alpha$, df1, df2), where df1 are the degrees of freedom of the numerator and df2 are the degrees of freedom of the denominator. Given an $F$ statistic calculated by equation 2.18, in repeated experiments on normally distributed systems with equal population standard deviations, an $F$ ratio would equal or exceed $s_1^2/s_2^2$ with probability =FDIST($F$, df1, df2).

Table 2.5. Values of the critical two-tailed Grubbs's G statistic at 95% and 99% probability for a pair of outliers at either end of a data set (Grubbs 1950)

| $n$ | $G_{0.05",n}$ | $G_{0.01",n}$ |
|-----|---------------|---------------|
| 4   | 0.0002        | 0.0000        |
| 5   | 0.009         | 0.0018        |
| 6   | 0.0349        | 0.0116        |
| 7   | 0.0708        | 0.0308        |
| 8   | 0.1101        | 0.0563        |
| 9   | 0.1493        | 0.0851        |
| 10  | 0.1864        | 0.1150        |
| 11  | 0.2213        | 0.1148        |
| 12  | 0.2537        | 0.1738        |
| 13  | 0.2836        | 0.2016        |
| 14  | 0.3112        | 0.2280        |
| 15  | 0.3367        | 0.2530        |

The null hypothesis that there are no outliers is rejected if $g < G_{critical}$ where

$$g(\text{high pair}) = \frac{\sum_{i=1}^{i=n-2}\left(x_i - \bar{x}_{1...n-2}\right)^2}{\sum_{i=1}^{i=n}\left(x_i - \bar{x}_{1...n}\right)^2} \quad \text{or} \quad g(\text{low pair}) = \frac{\sum_{i=3}^{i=n}\left(x_i - \bar{x}_{3...n}\right)^2}{\sum_{i=1}^{i=n}\left(x_i - \bar{x}_{1...n}\right)^2}$$

### 2.4.5.2  Chi-square Test of a Variance against a Given Variance

The purpose of the $F$ test is to answer the question of whether data with two different sample variances might have come from a single population. The test does not tell one what the population variance might be. Given a value for $s^2$, the population variance, a sample variance ($s^2$ from $n$ measurements) might be tested against it using a chi-square test:

$$T = \frac{(n-1)s^2}{\sigma^2} \tag{2.19}$$

For a two-sided test, the null hypothesis, $H_0$, is that the variance of the population from which the data giving $s^2$ is drawn is equal to $\sigma^2$, and the alternative hypothesis is that it is not equal. $H_0$ is rejected at the 95% level if $T > \chi^2_{0.025,n-1}$ or $T < \chi^2_{0.975,n-1}$. In Excel the probability of a particular value of chi-square is given by =CHIDIST($\chi^2$, df), and the critical value of chi-square is =CHIINV($\alpha$, df) for probability $\alpha$ and df degrees of freedom.

### 2.4.5.3 Cochran Test for Homogeneity
of Variances

If a sample has been analyzed by $k$ laboratories $n$ times each, the sample variances of the $n$ results from each laboratory can be tested for homogeneity—that is, any variance outliers among the laboratories can be detected. The ISO-recommended test is the Cochran test. The statistic that is tested is

$$c = \frac{s^2_{max}}{\sum\limits_{i=1}^{i=k} s^2_i} \qquad (2.20)$$

where $s^2_{max}$ is the greatest laboratory variance and the denominator is the sum of all the variances. If $c > C_{critical}$ (see table 2.6), then there are grounds for rejecting the results. It is usual to test the variances first, then to test any outliers using the Grubbs's tests described above. Note that if $n = 2$ (i.e., each laboratory only makes two measurements), equation 2.20 is used with $s$ equal to the difference between the results.

Following are some data from an Australian soil trial in which a sample of soil was analyzed three times by each of 10 laboratories for TPA (total peroxide acidity expressed as moles per tonne). The results are given in spreadsheet 2.3. The sample variance is calculated as the square of the sample standard deviation, and the Cochran statistic is calculated from equation 2.20.

The Cochran critical values for 99% and 95% for 9 degrees of freedom are 0.4775 and 0.5727, respectively. The calculated value for laboratory A is 0.514, and so the laboratory fails at 95% but not at 99% (the actual probability is 98.5%). In this interlaboratory trial the laboratory would be flagged, but its data still included in calculations of group means and standard deviation if no other problems were encountered.

## 2.4.6  Testing Means

### 2.4.6.1  Testing against a Given Value

Having repeated a measurement, how can the result be compared with a given value? We have already encountered this scenario with tests against a regulatory limit, and the problem has been solved by looking at the result and its 95% confidence interval in relation to the limit (see figure 2.5). This procedure can be turned into a test by some simple algebra on the equation for the 95% confidence interval (equation 2.13). If the population mean, $\mu$ is to be found in the 95% confidence interval, in 95% of repeated measurements,

$$\mu = \bar{x} \pm \frac{t_{\alpha'',n-1}s}{\sqrt{n}} \qquad (2.21)$$

then a Student's $t$ can be defined,

Table 2.6. Critical 95% and 99% values of the Cochran statistic for *k* laboratories
each repeating the sample measurement *n* times

| | 95% critical values ($\alpha$ = 0.05) | | | 99% critical values ($\alpha$ = 0.01) | | |
|---|---|---|---|---|---|---|
| | *n* | | | *n* | | |
| *k* | 2 | 3 | 4 | 2 | 3 | 4 |
| 3 | 0.9669 | 0.8709 | 0.7978 | 0.9933 | 0.9423 | 0.8832 |
| 4 | 0.9064 | 0.7679 | 0.6839 | 0.9675 | 0.8643 | 0.7816 |
| 5 | 0.8411 | 0.6838 | 0.5981 | 0.9277 | 0.7885 | 0.6958 |
| 6 | 0.7806 | 0.6161 | 0.5321 | 0.8826 | 0.7218 | 0.6259 |
| 7 | 0.7269 | 0.5612 | 0.48 | 0.8373 | 0.6644 | 0.5687 |
| 8 | 0.6797 | 0.5157 | 0.4377 | 0.7941 | 0.6152 | 0.5211 |
| 9 | 0.6383 | 0.4775 | 0.4028 | 0.7538 | 0.5727 | 0.4812 |
| 10 | 0.6018 | 0.445 | 0.3734 | 0.7169 | 0.5358 | 0.4471 |
| 11 | 0.5696 | 0.4169 | 0.3482 | 0.683 | 0.5036 | 0.4207 |
| 12 | 0.5408 | 0.3924 | 0.3265 | 0.652 | 0.4751 | 0.3922 |
| 13 | 0.515 | 0.3709 | 0.3075 | 0.6236 | 0.4498 | 0.3698 |
| 14 | 0.4917 | 0.3517 | 0.2907 | 0.5976 | 0.4272 | 0.3499 |
| 15 | 0.4706 | 0.3346 | 0.2758 | 0.5737 | 0.4069 | 0.3321 |
| 16 | 0.4514 | 0.3192 | 0.2625 | 0.5516 | 0.3885 | 0.3162 |
| 17 | 0.4339 | 0.3053 | 0.2505 | 0.5313 | 0.3718 | 0.3018 |
| 18 | 0.4178 | 0.2927 | 0.2396 | 0.5124 | 0.3566 | 0.2887 |
| 19 | 0.4029 | 0.2811 | 0.2296 | 0.4949 | 0.3426 | 0.2767 |
| 20 | 0.3892 | 0.2705 | 0.2206 | 0.4786 | 0.3297 | 0.2657 |
| 21 | 0.3764 | 0.2607 | 0.2122 | 0.4634 | 0.3178 | 0.2557 |
| 22 | 0.3645 | 0.2516 | 0.2046 | 0.4492 | 0.3068 | 0.2462 |
| 23 | 0.3535 | 0.2432 | 0.1974 | 0.4358 | 0.2966 | 0.2376 |
| 24 | 0.3431 | 0.2354 | 0.1908 | 0.4233 | 0.2871 | 0.2296 |
| 25 | 0.3334 | 0.2281 | 0.1847 | 0.4115 | 0.2782 | 0.2222 |

The null hypothesis of homogeneity of variance is rejected if $c > C_{critical}$.

$$t = \frac{|\bar{x} - \mu|\sqrt{n}}{s} \tag{2.22}$$

which can be compared with the 95% value. So if $t > t_{\alpha'',n-1}$, the null hypothesis that $\mu$ is within the 95% confidence interval of the measurement in 95% of repeated experiments is rejected. Alternatively, the probability associated with the calculated $t$ can be obtained in Excel (=TDIST(t,n - 1,2)) and an appropriate decision made.

For example, a trace element solution certified reference material was analyzed seven times for chromium with results 0.023, 0.025, 0.021, 0.024, 0.023, 0.022, and 0.024 ppm. The solution was certified at 0.0248 ppm. The mean of the seven replicates is 0.0231 ppm, and the sample standard deviation is 0.0013 ppm. Using equation 2.22 with $\mu$ = 0.0248 ppm gives

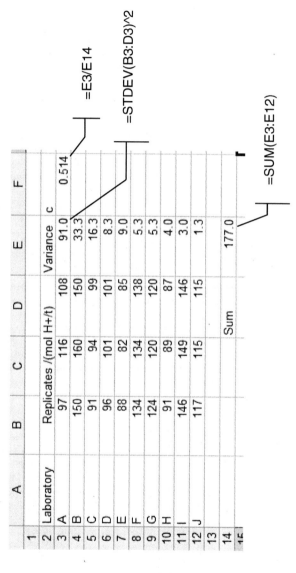

| | A | B | C | D | E | F |
|---|---|---|---|---|---|---|
| 1 | | | | | | |
| 2 | Laboratory | Replicates /(mol H+/t) | | | Variance | c |
| 3 | A | 97 | 116 | 108 | 91.0 | 0.514 |
| 4 | B | 150 | 160 | 150 | 33.3 | |
| 5 | C | 91 | 94 | 99 | 16.3 | |
| 6 | D | 96 | 101 | 101 | 8.3 | |
| 7 | E | 88 | 82 | 85 | 9.0 | |
| 8 | F | 134 | 134 | 138 | 5.3 | |
| 9 | G | 124 | 120 | 120 | 5.3 | |
| 10 | H | 91 | 89 | 87 | 4.0 | |
| 11 | I | 146 | 149 | 146 | 3.0 | |
| 12 | J | 117 | 115 | 115 | 1.3 | |
| 13 | | | | | | |
| 14 | | | | Sum | 177.0 | |
| 15 | | | | | | |

=E3/E14

=STDEV(B3:D3)^2

=SUM(E3:E12)

**Spreadsheet 2.3.** Cochran test for homogeneity of variances in the soil analysis example given in the text.

$$t = \frac{|0.0248 - 0.0231|\sqrt{7}}{0.0013} = 3.26$$

The 95% critical $t$ value for 6 degrees of freedom is 2.45. The calculated $t$ has a probability of 0.027. This is less than 0.05 (or $t = 3.26 > 2.45$), and so the null hypothesis that the bias is not significant is rejected, and we conclude that the results should be corrected for a bias of $0.0231 - 0.0248 = -0.0017$ ppm. The 95% confidence interval of this bias is $0.0013 \times 2.45/\sqrt{7} = \pm 0.0012$ ppm, which should be included in any measurement uncertainty estimate of the corrected results.

Note that no allowance has been made for the uncertainty of the certified reference material. Analysts often assume that the value of the reference material has been determined with much greater precision than their experiments. Suppose the certificate does give an expanded uncertainty of 0.0004 ppm (95% probability). The standard uncertainty is $u = 0.0004/2 = 0.0002$ ppm with, in the absence of any other information, infinite degrees of freedom. The $t$ value can now be recalculated as

$$t = \frac{|\bar{x} - \mu|}{\sqrt{u^2 + \dfrac{s^2}{n}}} = \frac{|0.0248 - 0.0231|}{\sqrt{0.0002^2 + \dfrac{0.0013^2}{7}}} = 2.81 \tag{2.23}$$

Although the $t$ is still significant ($2.81 > 2.45$), it is not as significant now that the extra source of uncertainty is included.

### 2.4.6.2 Testing Two Means

In testing it is common for a measurement to be made under two different conditions (different analysts, methods, instruments, times), and we must know if there is any significant difference in the measurements. In chapter 3 I discuss how ANOVA generalizes this problem and can compare a number of different factors operating on a result, but here I show that a $t$ test provides a reasonably straightforward test of two means. There are two formulas that can be used to provide the $t$ value, depending on whether it is assumed that the variances of the two populations from which the means are drawn are equal. Although an $F$ test (see 2.4.5.1 above) is a way of giving statistical evidence of the equality of the population variances, some people suggest always assuming unequal variances, as this is the more stringent test (i.e., it is more likely not to reject $H_0$). However, if two similar methods are being compared, it seems reasonable to assume that the precisions are likely to be the same. In this case the standard deviations are pooled, combining variances weighted by the number of degrees of freedom of each variance. That is, if one set of data has more observations, it contributes more to the pooled variance:

$$s_p = \sqrt{\frac{(n_1 - 1)s_1^2 + (n_2 - 1)s_2^2}{n_1 + n_2 - 2}} \qquad (2.24)$$

The $t$ value is then calculated

$$t = \frac{|\bar{x}_1 - \bar{x}_2|}{s_p \sqrt{1/n_1 + 1/n_2}} \qquad (2.25)$$

If the population variances are considered not to be equal, the $t$ value is calculated by

$$t = \frac{|\bar{x}_1 - \bar{x}_2|}{\sqrt{s_1^2/n_1 + s_2^2/n_2}} \qquad (2.26)$$

When the $t$ value is tested against a critical value, or the probability is calculated, the degrees of freedom is

$$df = \frac{\left(s_1^2/n_1 + s_2^2/n_2\right)^2}{\left(\dfrac{s_1^4}{n_1^2(n_1 - 1)} + \dfrac{s_2^4}{n_2^2(n_2 - 1)}\right)} \qquad (2.27)$$

rounded down to the nearest integer. Use the Excel function =ROUNDDOWN(df, 0).

### 2.4.6.3 Paired t Test

When two factors are being compared (e.g., two analysts, two methods, two certified reference materials) but the data are measurements of a series of independent test items that have been analyzed only once with each instance of the factor, it would not be sensible to calculate the mean of each set of data. However, the two measurements for each independent test item should be from the same population if the null hypothesis is accepted. Therefore the population mean of the *differences* between the measurements should be zero, and so the sample mean of the differences can be tested against this expected value (0). The arrangement of the data is shown in table 2.7.

The calculation of $t$ uses equation 2.22 with $\mu = 0$.

### 2.4.7 Student's t Tests in Excel

There are three ways to perform a $t$ test in Excel. First, a $t$ value can be calculated from the appropriate equation and then the probability calculated from =TDIST(t,df,tails), where df = degrees of freedom and tails = 2 or

Table 2.7. Arrangement of data and calculations for a paired $t$ test

| Test item | Result by method A | Result by method B | Difference, A − B |
|---|---|---|---|
| 1 | $x_{A,1}$ | $x_{B,1}$ | $d_1 = x_{A,1} - x_{B,1}$ |
| 2 | $x_{A,2}$ | $x_{B,2}$ | $d_2 = x_{A,2} - x_{B,2}$ |
| 3 | $x_{A,3}$ | $x_{B,3}$ | $d_3 = x_{A,3} - x_{B,3}$ |
| 4 | $x_{A,4}$ | $x_{B,4}$ | $d_4 = x_{A,4} - x_{B,4}$ |
| ... $n$ | $x_{A,n}$ | $x_{B,\bullet}$ | $d_n = x_{A,n} - x_{B,n}$ |
| | | Mean | $\bar{d} = \dfrac{\sum\limits_{i=1}^{i=n} d_i}{n}$ |
| | | Sample standard deviation | $s_d = \sqrt{\dfrac{\sum\limits_{i=1}^{i=n}(d_i - \bar{d})^2}{n-1}}$ |
| | | $t$ | $t = \dfrac{|\bar{d}|\sqrt{n}}{s_d}$ |

Measurements by method A are compared with those by method B. Each item is different and is measured once by each method.

1 for a two-tailed or one-tailed test. Alternatively, the $t$ value can be compared with a critical $t$ value =TINV($\alpha$, df), where $\alpha = 0.05$ for a 95% test, or 0.01 for a 99% test. Excel also provides a function =TTEST(range 1, range 2, tails, type). The two ranges contain the data, as before tails is 1 or 2, and type = 1 for a paired $t$ test (in which case the lengths of the two data ranges must be the same), = 2 for a means test with the assumption of equal variance (using equation 2.25), and = 3 for a means test with assumption of unequal variance (equation 2.26). The result is the probability associated with the $t$-value.

Finally, the "Add Ins ..." of the Data Analysis Tools gives the three methods (paired, means with equal variance, and means with unequal variance) as menu-driven options (see spreadsheet 2.4). Access Data Analysis ... from the Tools menu (if it is not there, the Analysis ToolPak needs to be installed via the Add-Ins ... menu, also found in the Tools menu).

## 2.5 Analysis of Variance

The workhorse of testing, ANOVA, allows the variance of a set of data to be partitioned between the different effects that have been allowed to vary during the experiments. Suppose an experiment is duplicated at combinations of different temperatures and different pH values. ANOVA can calculate the variances arising from random measurement error, from changing the temperature, and from changing the pH. The significance of temperature and pH effects is assessed by comparing the variance attributable to each effect with that of the measurement. (A significant effect is one that is greater than would be expected from random measurement error alone.) ANOVA only works on normally distributed data, but its ability to cope with any number of factors makes it well used.

The idea behind ANOVA is that if two distributions with very different means are combined, the variance of the resulting distribution will be much greater than the variances of the two distributions. Here I describe one- and two-way ANOVA in Excel. See (Massart et al 1997, chapter 6) for a more comprehensive discussion of the use of ANOVA.

### 2.5.1   One-Factor ANOVA

One-factor, or one-way, ANOVA considers replicate measurements when only one factor is being changed. Unlike a *t* test that can only accommodate two instances of the factor, ANOVA is generalized for any number. Here "factor" means the kind of thing being studied (method, temperature setting, instrument, batch, and so on), whereas an instance of a factor is the particular value (20°C, 30°C, 40°C; or types of chromatography or enzymatic methods). Do not confuse factor and instance of a factor. A one-way ANOVA can be used to analyze tests of three analysts analyzing a portion of test material five times. The factor is "analyst." and the instances of the factor are "Jim," "Mary," and "Anne." In this example three *t* tests could be done between the results from Jim and Mary, Jim and Anne, and Mary and Anne, but ANOVA can first tell if there is any significant difference among the results of the three, and then, using the method of least significant differences, reveal which analyst is different from the other two, or if all three are different from each other.

#### 2.5.1.1   *Data Layout and Output*
####             *in One-Factor ANOVA*

The data are laid out in adjacent columns with a suitably descriptive header. Using an example of the analysis of glucose in a sample by the three analysts, the data would be entered in a spreadsheet as shown in spreadsheet 2.5.

In Excel, there are three options for ANOVA: one factor, two factor without replication, and two factor with replication. The different options in the Data Analysis Tools menu are shown in spreadsheet 2.6.

Choosing the one-factor option (which must have replicate data, although not necessarily the same number of repeats for each instance of the factor), the output includes an ANOVA table shown in spreadsheet 2.5. In the output table, SS stands for sum of squares, df is degrees of freedom, and MS is

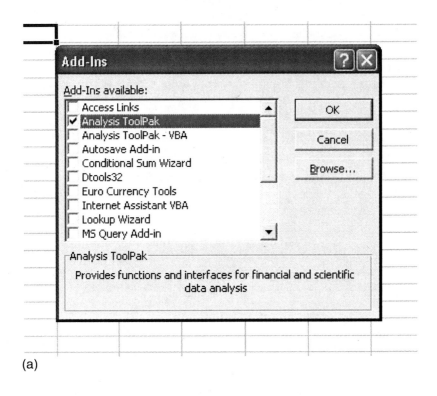

(a)

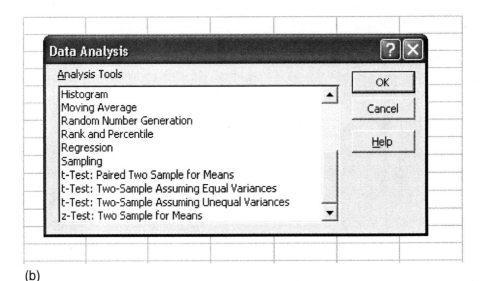

(b)

**Spreadsheet 2.4.** (a) The Add-In menu. (b) The Data Analysis Tools menu showing the three *t* test options. (c) Menu for the *t* test assuming equal variances.

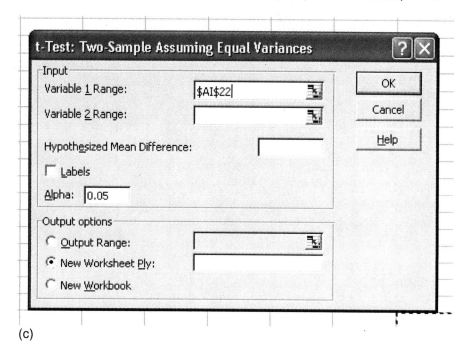

(c)

the mean square. $F$ is the Fisher $F$ value, calculated as the ratio of between-groups MS and within-groups MS. The degrees of freedom for the between groups is $k-1$, where there are $k$ instances of the factor (number of columns of data), and the degrees of freedom for the whole data set is $N-1$, where $N$ is the total number of data. The difference, $N-k$, is the degrees of freedom within the groups. The within-groups mean square is the pooled variances of the instances of the factor and is usually equated with the random measurement variance. When calculations are done using all the data, and going across the columns, the variance is now a combination of the random measurement error (which never goes away) and variance arising from the differences between the instances (here Jim, Mary, and Anne). So the question, is there a significant difference between the analysts can be answered by determining if the between-groups MS (combination of measurement variance and the difference between the analysts) is significantly greater than the within-groups MS (the measurement variance). This determination can be made by using a one-tailed $F$ test of the ratio of the variances. The test is one tailed because the way the problem is set up ensures that the between-groups MS is greater than the within-groups MS (both contain the same measurement variance, and the between-groups MS has the extra component due to the differences between the instances of the factor). The probability ($P$) is the one-tailed probability that a greater $F$ statistic would be found for a repeated experiment if the null hypothesis that there is no difference between the analysts were true. As the value of $10^{-5}$ is small, the null

### (a)

| Jim | Mary | Anne |
|------|------|------|
| 1.24 | 1.25 | 1.21 |
| 1.26 | 1.27 | 1.20 |
| 1.27 | 1.26 | 1.23 |
| 1.26 | 1.27 | 1.22 |
|      | 1.28 | 1.21 |
|      |      | 1.21 |

### (b)

Anova: Single Factor

SUMMARY

| Groups | Count | Sum | Average | Variance |
|--------|-------|------|----------|----------|
| Jim | 4 | 5.03 | 1.2575 | 0.000158 |
| Mary | 5 | 6.33 | 1.266 | 0.00013 |
| Anne | 6 | 7.28 | 1.213333 | 0.000107 |

ANOVA

| Source of Variation | SS | df | MS | F | P-value | F crit |
|---------------------|------|------|------|------|----------|--------|
| Between Groups | 0.008765 | 2 | 0.004383 | 34.41003 | 1.07E-05 | 3.88529 |
| Within Groups | 0.001528 | 12 | 0.000127 | | | |
| | | | | | | |
| Total | 0.010293 | 14 | | | | |

**Spreadsheet 2.5.** (a) Input to and (b) output from the one-factor ANOVA example.

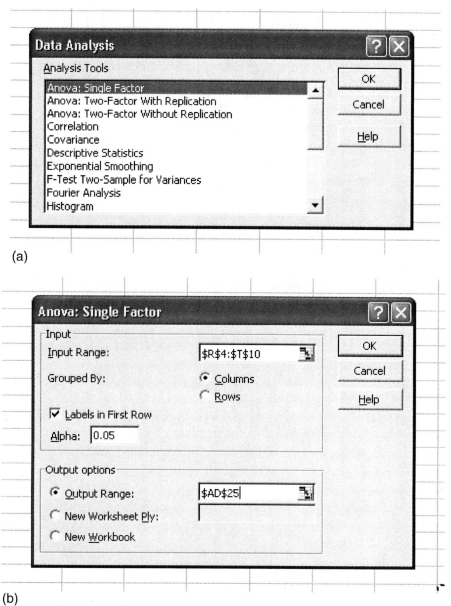

(a)

(b)

**Spreadsheet 2.6.** (a) Excel Data Analysis ToolPak menu showing ANOVA options. (b) Single-factor ANOVA menu.

hypothesis would be rejected. The critical 95% $F$ value is also given, and as the calculated $F$ exceeds it, $H_0$ would be rejected at this probability level. If the mean and 95% confidence interval of each person is plotted (figure 2.9), the fact that Anne is obtaining smaller values than the other two analysts becomes apparent. An ANOVA with the data from Jim and Mary shows no significant difference between them. A $t$ test of Jim and Mary under the assumption of equal variances gives the same result.

### 2.5.1.2  Calculating Variances

Sometimes knowing that there is a significant difference between the groups is all that is needed. In the example of the analysts, quantifying the variances is not particularly useful. However, if the factor were pH, it might be of interest to know the variance associated with the change in this factor. Remember that the within-groups mean square is an estimate of the measurement variance ($\sigma^2$). If each of the instances has the same number of observations ($n$), then the between-groups mean square estimates $\sigma^2 + n\sigma_f^2$, where $\sigma_f^2$ is the variance due to the factor. With different numbers of observations, the between-groups mean square estimates $\sigma^2 + n'\sigma_f^2$, where $n'$ is calculated as

$$n' = \frac{N^2 - \sum_{i=1}^{i=k} n_i^2}{(k-1)N} \qquad (2.28)$$

where $n_i$ is the number of observations for the $i$th factor instance. This equation is not as complex as it looks; $N = \sum_{i=1}^{i=k} n_i$ is the total number of observations, and $k$ is the number of instances of the factor. In the example, $k = 3$, $N = 15$, and $n'$ is

$$n' = \frac{225 - 77}{2 \times 15} = 4.93$$

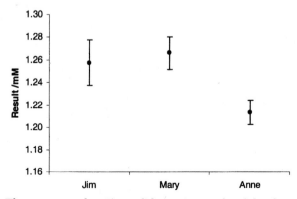

**Figure 2.9.** The means and 95% confidence intervals of the data shown in spreadsheet 2.5.

The $n'$ will always fall between the minimum and maximum numbers of observations. For the analyst example, the between-groups mean square = 0.004383, the within-groups mean square = 0.000127, and $n' = 4.93$. Therefore, the measurement standard deviation $(\sigma) = \sqrt{(0.000127)} = 0.011$ mM, and the between-analysts standard deviation $(\sigma_f) = \sqrt{[(0.004383 - 0.000127)/4.93]} = 0.029$ mM.

### 2.5.2  Two-Factor ANOVA

Excel can compute an ANOVA for two factors if there is a single measurement at each combination of the factor instances, or if each combination has been replicated the same number of times. In the example above, if the analysts are given a different batch of enzyme and are asked to perform duplicate measurements, a two-way ANOVA can be performed. The two factors are "analyst" with three instances ("Jim," "Mary," and "Anne"), and "batch" with two instances ("batch 1" and "batch 2"). Spreadsheet 2.7 gives the data and output from the analysis, and spreadsheet 2.8 shows the Excel menu. With the two-factor with replication option the data in the input range must include the row and column labels. (This is why there is no box to check for labels, an option in the other ANOVA menus.) "Columns" refers to the factor in the columns (here the analysts) and "sample" refers to the rows (here batches). In addition, the output table has gained an extra row called "interactions." An interaction effect is one in which the variance associated with one factor depends on the value of another factor. If factors are independent, then the interaction effect is zero. As before, the ANOVA shows that the analysts factor contributed a significant variance $(P = 0.007795)$, but the batches of enzyme are not quite significant at the 95% level $(P = 0.06163)$, and the interaction is far below significance $(P = 0.5958)$. How should this be interpreted? Anne is still the odd person out in the analysts factor. The batches of enzyme are likely to be of different activity because of the way enzymes are prepared, so although the ANOVA tells us this difference is not significant at the 95% level (in fact, it is significant at the 93.8% level), it would be better to interpret this as a real difference. There is no reason to believe that the analysts' techniques and measurement procedure would depend on which batch of enzyme was used, and so it is not surprising that the interaction effect is small and not significant.

In the absence of replicate measurements, it is not possible to estimate interaction effects. These are bundled into the "error" term (see spreadsheets 2.9 and 2.10). Even with the sparse data of a single measurement for each combination of factors, the result is in agreement with the other calculations. The analysts are still significantly different ("columns" $P = 0.04452$) but the batches are not ("rows" $P = 0.09274$).

Because of the limitations of Excel, I will not explore the use of the ANOVA further, but with appropriate software, or with a calculator and much patience, the method can be applied to any number of factors.

(a)

| | Jim | Mary | Anne |
|---|---|---|---|
| Batch1 | 1.24 | 1.25 | 1.21 |
| | 1.26 | 1.27 | 1.20 |
| Batch2 | 1.25 | 1.26 | 1.23 |
| | 1.28 | 1.28 | 1.24 |

ANOVA

| Source of Variation | SS | df | MS | F | P-value | F crit |
|---|---|---|---|---|---|---|
| Sample | 0.001008 | 1 | 0.001008 | 5.26087 | 0.061634 | 5.987374 |
| Columns | 0.00465 | 2 | 0.002325 | 12.13043 | 0.007795 | 5.143249 |
| Interaction | 0.000217 | 2 | 0.000108 | 0.565217 | 0.595807 | 5.143249 |
| Within | 0.00115 | 6 | 0.000192 | | | |
| | | | | | | |
| Total | 0.007025 | 11 | | | | |

(b)

**Spreadsheet 2.7.** (a) Input data and (b) ANOVA table for two factors with replication.

**Spreadsheet 2.8.** Excel menu for two-factor ANOVA with replication.

## 2.6 Linear Calibration

Most analyses require some form of calibration. A typical instrument does not give, oraclelike, the answer to the analyst's question. An instrument provides what is known as an "indication of a measuring instrument" (ISO 1993b, term 3.1), and then that indication must be related to the concentration or to whatever is being measured. Whether the measurements are peak heights, areas, absorbances, counts of a mass spectrometer detector, or nuclear magnetic resonance peak areas, each of these is related to the amount (or amount concentration) of the species being detected by a calibration function. Typically, although increasingly not so, this function is linear in the concentration:

$$y = a + bx \qquad (2.29)$$

where the indication is $y$, the concentration is $x$, and $a$ and $b$ are parameters of the calibration model. From days when these relationships were established by a graph, $a$ is known as the intercept and $b$ is the slope.

### 2.6.1 Slopes and Intercepts

The values of $a$ and $b$ are estimated by observing the indication for a series of standard samples for which the concentrations are known. Assuming that

|  | Jim | Mary | Anne |
|---|---|---|---|
| Batch1 | 1.25 | 1.26 | 1.205 |
| Batch2 | 1.265 | 1.27 | 1.235 |

(a)

ANOVA

| Source of Variation | SS | df | MS | F | P-value | F crit |
|---|---|---|---|---|---|---|
| Rows | 0.000504 | 1 | 0.000504 | 9.307692 | 0.092735 | 18.51276 |
| Columns | 0.002325 | 2 | 0.001162 | 21.46154 | 0.044521 | 19.00003 |
| Error | 0.000108 | 2 | 5.42E-05 | | | |
| | | | | | | |
| Total | 0.002937 | 5 | | | | |

(b)

**Spreadsheet 2.9.** (a) Input data and (b) ANOVA table for two factors without replication.

**Spreadsheet 2.10.** Excel menu for two-factor ANOVA without replication.

(1) all the variance is in $y$, (2) the variance is normally distributed and independent of concentration, and (3) the measurement model is correct, then a classical least squares gives us the estimates of the parameters:

$$\hat{b} = \frac{\sum_{i=1}^{i=n}\left[(x_i - \bar{x})(y_i - \bar{y})\right]}{\sum_{i=1}^{i=n}(x_i - \bar{x})^2} \qquad (2.30)$$

$$\hat{a} = \bar{y} - \hat{b}\bar{x} \qquad (2.31)$$

where $\bar{x}$ and $\bar{y}$ are the means of the calibration data, and the model that is fitted is

$$y = \hat{a} + \hat{b}x + \varepsilon \qquad (2.32)$$

and $\varepsilon$ is the random error in y with mean zero. The hats (^) on $a$ and $b$ in equations 2.30–2.32 denote that these values are estimates, but from this point they will be omitted. Calibration is best done with a short range of concentrations and as many points as possible. The goodness of fit is measured by the standard error of the regression (not the correlation coefficient), $s_{y/x}$, given by

$$S_{y/x} = \sqrt{\frac{\sum\limits_{i=1}^{i=n}(y_i - \bar{y})^2}{df}}$$

$$(2.33)$$

where $\bar{y}$ is the estimated value from the measurement model $\bar{y} = a + bx$ and differs from the value of $y$ by $\varepsilon$. The degrees of freedom is $n - 2$, or for a model that is forced through zero ($a \equiv (0)$, $n - 1$, where there are $n$ points in the calibration. If it is necessary to know the standard errors (standard deviations) of the slope and intercept, these are calculated as:

$$S_b = \frac{S_{y/x}}{\sqrt{\sum\limits_{i=1}^{i=n}(x_i - \bar{x})^2}}$$

$$(2.34)$$

$$S_a = S_{y/x}\sqrt{\frac{\sum\limits_{i}x_i^2}{n\sum\limits_{i}(x_i - \bar{x})^2}}$$

$$(2.35)$$

A 95% confidence interval on slope or intercept is obtained by multiplying these standard errors by a $t$ value at the desired probability and $n - 2$ ($n - 1$) degrees of freedom:

$$b \pm t_{\alpha'',df}S_b$$

$$(2.36)$$

$$a \pm t_{\alpha'',df}S_a$$

$$(2.37)$$

All these statistics are available in Excel in the LINEST function. This gives an array output, and for fitting data to a straight line (equation 2.29) is created by following these steps:

1. Choose an empty two column by five row block and select it.
2. Type =LINEST(y-range,x-range,1,1), in the function menu bar, where x-range are the cells containing the $x$ values, and y-range are the cells containing the $y$ values
3. Press Ctrl-Shift-Enter. (If only one cell appears, then only Enter has been pressed, or the block was not selected).

The output contains the information shown in table 2.8. The values of slope and intercept are in the first row, and their standard errors are below these in the second row.

### 2.6.1.1  Forcing through the Origin

If the analysis is corrected for background or blanks, there is often a presumption that the calibration line will pass through the origin. Thus the calibration is

Table 2.8. Output of the Excel LINEST function, =LINEST(y-range,x-range,1,1)

| Slope: $b$ | Intercept: $a$ |
|---|---|
| Standard deviation of slope: $s_b$ | Standard deviation of intercept: $s_a$ |
| Coefficient of determination: $r^2$ | Standard error of regression: $s_{y/x}$ |
| Fisher $F$ statistic: $F = \dfrac{\overline{SS}_{\text{regression}}}{\overline{SS}_{\text{residual}}}$ | Degrees of freedom of the regression: $df = n - 1$ or $n - 2$ |
| Sum of squares due to regression: $SS_{\text{regression}}$ | Sum of squares due to residual: $SS_{\text{residual}}$ |

$$(y - y_{\text{blank}}) = bx \tag{2.38}$$

and a concentration that is calculated from equation 2.38 is

$$\bar{x} = \frac{(y - y_{\text{blank}})}{b} \tag{2.39}$$

and $y_{\text{blank}}$ stands in for the calculated intercept, $a$. The good news is that a degree of freedom is gained, only calculating one parameter (the slope) from the calibration data. However, there is a risk that if $y_{\text{blank}}$ is not a true indication of the tendency of the calibration at $x = 0$, there still might be a non-zero intercept. As part of the method validation, and from time to time, the calibration data should be fitted to the model with an intercept (equation 2.29), and the 95% confidence interval of the intercept shown to include zero. This is important because even a small nonzero intercept can lead to huge errors if it is not noticed and the line is forced through zero.

### 2.6.1.2   One-Point and Two-Point Calibration

A single measurement of a calibration sample can give the concentration of the test solution by a simple ratio. This is often done in techniques where a calibration internal standard can be measured simultaneously (within one spectrum or chromatogram) with the analyte and the system is sufficiently well behaved for the proportionality to be maintained. Examples are in quantitative nuclear magnetic resonance with an internal proton standard added to the test solution, or in isotope dilution mass spectrometry where an isotope standard gives the reference signal. For instrument responses $I_{\text{IS}}$ and $I_{\text{sample}}$ for internal standard and sample, respectively, and if the concentration of the internal standard is $c_{\text{IS}}$, then

$$c_{\text{sample}} = \frac{I_{\text{sample}} c_{\text{IS}}}{I_{\text{IS}}} \tag{2.40}$$

If the concentrations of sample and internal standard are carefully matched, then the method can be very accurate (Kemp 1984).

A variation of this technique that gives a bit more information is to bracket the concentration of the test sample by two calibration samples ($c_1$ and $c_2$, giving instrument responses $I_1$ and $I_2$), when

$$C_{sample} = c_1 + \frac{c_2 - c_1}{I_2 - I_1} \times \left(I_{sample} - I_1\right) \tag{2.41}$$

Neither one-point nor two-point calibrations have room to test the model or statistical assumptions, but as long as the model has been rigorously validated its use in the laboratory has been verified, these methods can work well. Typical use of this calibration is in process control in pharmaceutical companies, where the system is very well known and controlled and there are sufficient quality control samples to ensure on-going performance.

### 2.6.2  Estimates from Calibrations

The purpose of a calibration line is to use it to estimate the concentration of an unknown sample when it is presented to the instrument. This is achieved by inverting the calibration equation to make $x$ the subject. For an indication $y_0$,

$$\hat{x} = \frac{y_0 - \hat{a}}{\hat{b}} \tag{2.42}$$

The uncertainty of this estimate is

$$s_{\hat{x}_0} = \frac{s_{y/x}}{\hat{b}} \sqrt{\frac{1}{m} + \frac{1}{n} + \frac{\left(y_0 - \bar{y}\right)^2}{\hat{b}^2 \sum\limits_{i=1}^{i=n}\left(x_i - \bar{x}\right)^2}} \tag{2.43}$$

where the indication of the unknown solution has been observed $m$ times (if $m$ is greater than 1, $y_0$ is the average of the $m$ observations), and other coefficients and terms have been defined above. Multiplication by the 95%, two-tailed $t$ value at $n - 2$ degrees of freedom gives the 95% confidence interval on the estimate:

$$\hat{x} \pm t_{\alpha'',n-2} s_{\hat{x}_0} \tag{2.44}$$

This confidence interval arises from the variance in the calibration line and instrument response to the test sample only. Any other uncertainties must be combined to give the overall confidence interval (expanded uncertainty). The form of equation 2.43 is instructive. The standard deviation of the estimate is reduced by (1) having more points in the calibration line (greater $n$), (2) repeating the unknown more times (greater $m$), (3) having a greater sensitivity (slope), (4) having a better fit (smaller $s_{y/x}$) and (5) performing the measurement in the middle of the calibration ($y_0 - \bar{y}$) = 0. The effect of the number of points in the calibration line is twofold. First, $n$ appears in equation 2.43 as $1/n$, but more important for small $n$ (say < 10), it sets the $t$ value in equation 2.44 through the $n - 2$ degrees of freedom. The slope of the calibration, $b$, is also known as the *sensitivity* of the calibration. A greater sensitivity (i.e., when the instrument reading increases rapidly with concentration), leads to a more precise result. When validating a method (see chapter 8) these calibration parameters are established, together with the detection limit and range of linearity.

## Note

1. "Student" was the pseudonym of W.S. Gosset, brewing chemist and part-time statistician. Now the $t$ value may be referred to without "Student," but I like to mention its originator.

## References

Grubbs, F E (1950), Sample criteria for testing outlying observations. *The Annals of Mathematical Statistics,* 21 (1), 27–58.

Hibbert, D B and Gooding, J J (2005), *Data analysis for chemistry* (New York: Oxford University Press).

ISO (1994), Precision of test methods—Part 1: General principles and definitions, 5725-1 (Geneva: International Organization for Standardization).

ISO (1993a), *Guide to the expression of uncertainty in measurement,* 1st ed. (Geneva: International Organization for Standardization).

ISO (1993b), *International vocabulary of basic and general terms in metrology,* 2nd ed. (Geneva: International Organization for Standardization).

Joint Committee for Guides in Metrology (2007), *International vocabulary of basic and general terms in metrology,* 3rd ed. (Geneva: International Organization for Standardization).

Kemp, G J (1984), Theoretical aspects of one-point calibration: causes and effects of some potential errors, and their dependence on concentration. *Clinical Chemistry,* 30 (7), 1163–67.

Massart, D L, Vandeginste, B. G. M., Buydens, J. M. C., de Jong, S., Lewi, P. J. Smeyers-Verberke, J. (1997), *Handbook of chemometrics and qualimetrics part A,* 1st ed., vol 20A in *Data handling in science and technology* (Amsterdam: Elsevier).

Mullins, E (2003), *Statistics for the quality control chemistry laboratory,* 1st ed. (Cambridge: The Royal Society of Chemistry).

Oakland, J S (1992), *Statistical process control* (New York: John Wiley and Sons).

# 3

## Modeling and Optimizing Analytical Methods

### 3.1 Introduction

I asked a professor, visiting from a nation well regarded for its hardworking ethos, whether in his search for ever better catalysts for some synthesis or other, he used experimental design. His answer was, "I have many research students. They work very hard!" Many people believe that an infinite number of monkeys and typewriters would produce the works of Shakespeare, but these days few organizations have the luxury of great numbers of researchers tweaking processes at random in order to make them ever more efficient. The approach of experimental scientists is to systematically change aspects of a process until the results improve. In this chapter I look at this approach from a statistical viewpoint and show how a structured methodology, called experimental design, can save time and effort and arrive at the best (statistically defined) result. It may be a revelation to some readers that the tried-and-trusted "change one factor at a time" approach might yield incorrect results, after requiring more experiments than is necessary. In the sections that follow, I explain how experimental design entails more than just having an idea of what you are going to do before beginning an experiment.

## 3.2  Optimization in Chemical Analysis

### 3.2.1  What Is Optimization?

Optimization is the maximizing or minimizing a response by changing one or more input variables. In this chapter optimization is synonymous with maximization, as any minimization can be turned into a maximization by a straightforward transformation: Minimization of cost can be seen as maximization of profit; minimization of waste turns into maximization of production; minimization of $f(x)$ is maximization of $1/f(x)$ or $-f(x)$. Before describing methods of effecting such an optimization, the term optimization must be carefully defined, and what is being optimized must be clearly understood.

There are some texts on experimental design available for chemists, although often the subject is treated, as it is here, within a broader context. A good starter for the basics of factorial designs is the *Analytical Chemistry Open Learning* series (Morgan 1991). Reasonably comprehensive coverage is given in Massart et al.'s (1997) two-volume series, and also in a book from the Royal Society of Chemistry (Mullins 2003). If you are an organic chemist and want to optimize a synthesis, refer to Carlson and Carlson (2005). Experimental design based on the use of neural networks to guide drug discovery is treated by Zupan and Gasteiger (1999). For the historically minded, original papers by Fisher (1935) and, for Simplex optimization, Deming and Morgan (1973) should be consulted.

### 3.2.2  What Do You Want to Optimize?

Everyone wants to improve their existence. Many philosophies stress that a good life is one that leaves the world a better place than when that life began. So what is *better*? Before trying to make improvements, one must know exactly what is being improved and whether the improvement can be measured. This sounds simpler than it often is. A synthetic organic chemist usually wants to maximize yield of product, but would a sensible person spend an extra month going from 90% yield to 91%? Perhaps they would. Someone working for a major multinational company that improved the efficiency of the production of chlorine by 1% would save their company millions of dollars each year. The first message is that optimization lies very much in the eye of the beholder, and that whatever is being optimized, the bottom line is invariably time and cost. Indeed, if time is money, as Henry Ford observed, cost is only ever optimized.

The second message is that whatever is being optimized must be measurable. It must be possible objectively determine whether the endeavors of the scientist have been successful. To obtain the full benefit from optimization, the entity measured (called a "response") should have a clear mathematical relationship with the factors being studied to effect the optimization.

The approach I am describing comes from the great improvements in manufacturing that were wrought in the first half of the twentieth century by stalwarts such as Shewhart, Ishikawa, and Deming. They viewed the production of a factory in terms of inputs, the process, and outputs (figure 3.1).

By observing and measuring the output (cars, ball bearings, washing machines), the inputs (not just raw materials but personnel, energy, etc.), and the process (settings of machines and other variables) could be changed to effect a measurable improvement. Once the process was optimized, the output could be continuously monitored for noncompliance and feedback used to correct and prevent future noncompliance. An analytical experiment may be fitted into this view at two levels. At the highest, the systems level, a sample is the input, and by virtue of an analytical procedure a result is achieved as output. At a level that would be amenable to experimental design, a number of controllable factors relating to an instrumental measurement in the analytical procedure lead to observations made ("indications of the measuring instrument"). Changing the values of the factors causes the observations to change, and allowing for optimization of the values of the factors.

This is a very brief representation of a major subject, but in so far as the principles of process control and experimental design can be applied to analytical chemistry, suitable ideas and concepts will be plundered that will help the improvement of an analysis.

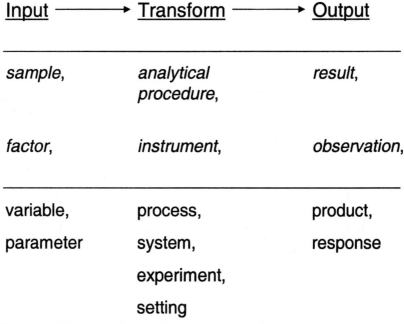

| Input | Transform | Output |
|---|---|---|
| sample, | analytical procedure, | result, |
| factor, | instrument, | observation, |
| variable, | process, | product, |
| parameter | system, | response |
| | experiment, | |
| | setting | |

**Figure 3.1.** The process view of chemical analysis.

What does optimization mean in an analytical chemical laboratory? The analyst can optimize responses such as the result of analysis of a standard against its certified value, precision, detection limit, throughput of the analysis, consumption of reagents, time spent by personnel, and overall cost. The factors that influence these potential responses are not always easy to define, and all these factors might not be amenable to the statistical methods described here. However, for precision, the sensitivity of the calibration relation, for example (slope of the calibration curve), would be an obvious candidate, as would the number of replicate measurements needed to achieve a target confidence interval. More examples of factors that have been optimized are given later in this chapter.

### 3.2.3  What Do You Change to Reach an Optimum?

#### 3.2.3.1  Factors

Having thought carefully about what should be optimized, your next task is to identify those factors that have some effect on the response. There is no point in changing the temperature of an experiment if temperature has no effect on the measured response. When doing experimental design, the first exercise is to perform experiments to identify which factors have a significant effect and that therefore will repay the effort of optimization. Sometimes this step is not required because the factors for optimization are obvious or are already given. Factors that influence the response are called "controlled" if there is some way of setting their values and "uncontrolled" if not. Whether a factor is controlled or uncontrolled depends on the experiment. With a water bath, temperature becomes controlled to about ± 0.5°C; in the open laboratory without climate control, a variation of 5° or 10°C might be expected, and temperature would be considered uncontrolled. There is always a range of factor values implied or explicit in any optimization. Temperature cannot go below absolute zero, and for many processes the practical limit is the temperature of liquid nitrogen or even an ice bath. The maximum temperature is constrained by available heating and by the physical and chemical limits of the experiment and apparatus.

The term "factor" is a catch-all for the concept of an identifiable property of a system whose quantity value might have some effect on the response. "Factor" tends to be used synonymously with the terms "variable" and "parameter," although each of these terms has a special meaning in some branches of science. In factor analysis, a multivariate method that decomposes a data matrix to identify independent variables that can reconstitute the observed data, the term "latent variable" or "latent factor" is used to identify factors of the model that are composites of input variables. A latent factor may not exist outside the mathematical model, and it might not therefore influence the response directly, but the concept is a convenient umbrella for a large number of factors that have an effect on the response.

Ranges of factors studied and the values investigated are often shown on a graph in $n$ dimensions in which the ranges of values of $n$ factors define a set of axes and points indicate the experimental values investigated. Such graphs are most useful for two or three factors because of the obvious limitations of drawing in multiple dimensions. The area or volume described is called "factor space," being the combinations of values of factors for which experiments can be performed (see figure 3.2).

A final point about factors. They need not be continuous random variables. A factor might be the detector used on a gas chromatograph, with values "flame ionization" or "electron capture." The effect of changing the factor no longer has quite the same interpretation, but it can be optimized—in this case simply by choosing the best detector.

### 3.2.3.2   Local and Global Optima

The optimum response is found within the factor space. Consider an $n + 1$ dimensional space in which the factor space is defined by $n$ axes, and the final dimension ($y$ in two dimensions, and $z$ in three) is the response. Any combination of factor values in the $n$ dimensional factor space has a response associated with it, which is plotted in the last dimension. In this space, if there are any optima, one optimum value of the response, called the "global optimum," defines the goal of the optimization. In addition, there may be any number of responses, each of which is, within a sublocality of the factor space, better than any other response. Such a value is a "local optimum."

A response can only be drawn as a function of one or two factors because of the constraints of the three-dimensional world, but real optimizations are often functions of tens of factors, which is why a more structured approach to optimization is necessary. Although the systems requiring optimization can be very complex, in reality experimental systems are rarely as multimodal as the purely mathematical example in figure 3.3b and c. The effect of a factor such as temperature, pressure, or the concentration of a reagent may not necessarily be linear, but it always has a single value (i.e., for a given set of factor values, there is only one response), and it rarely exhibits more than one maximum, and sometimes not even one.[1] Thus, across a chosen range, the response increases with a factor, and so the optimum response has more to do with determining the maximum factor value possible than it does with locating the position of a real maximum in the factor range. In figure 3.3a, a common situation is portrayed in which the response is more or less flat across a range of factor values. Here, the yield falls off at high and low pH, but around neutral the yield is acceptable. So, although there is a maximum that could be located by experiment, the indicated range is more than adequate, and time should not be wasted finding the ultimate optimum. Having discovered the shape of the response of figure 3.3a, other

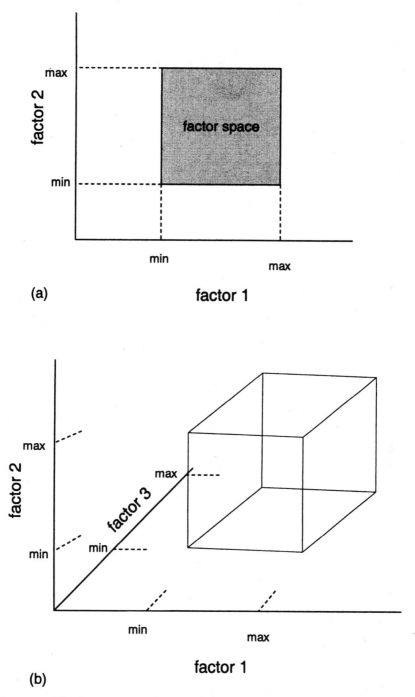

**Figure 3.2.** The factor space of two and three factors. The square (a) and cube (b) define the space within which combinations of factor values can be applied to experiments.

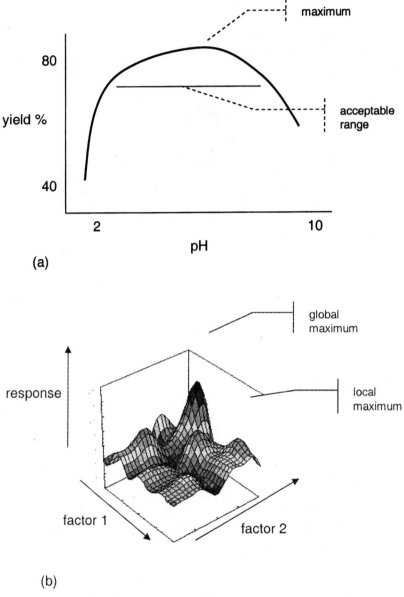

**Figure 3.3.** Examples of local and global optima. (a) One factor (pH) with the response (yield) giving a maximum, but with a wide acceptable range. (b) Function of two variables giving a number of maxima. The global maximum and one local maximum are shown. (c) The response surface of graph b as a contour map.

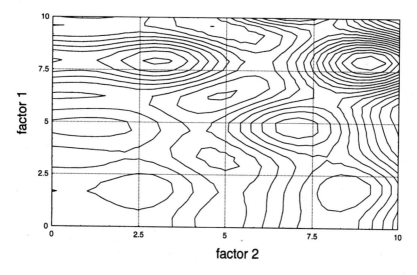

(c)

considerations may be used to choose a particular pH to run the experiment. Alternatively, these other considerations, perhaps the time or cost of the experiment, could be combined in the response to be optimized (e.g., yield × time) to give a more defined maximum. Remember that the shape of the response curve is rarely known, and the aim of the procedure is to find a reasonable value of the factor with a reasonable effort (i.e., low number of experiments). The experimenter must rely on the results of the experiments that have been performed.

### 3.2.3.3  Correlation between Factors

A two- or three-dimensional graph drawn with axes at right angles implies that each factor can take completely independent values and that a response exists for every point on the graph. This may be the case for many factors. If the effects of initial pH and time of an experiment are being studied, for example, it is probably acceptable to set the pH of the solution then allow the experiment to go for as long as required. However, if a simple water/methanol mobile phase for liquid chromatography is used, it would be a waste of an axis to try to optimize both water and methanol concentration, because what is not water is methanol and vice versa. This is only a one-factor problem—the fraction of water (or methanol).

Figure 3.4 shows the water–methanol mobile phase example of a mixture problem. A three-component mixture that must add up to 100% can be represented on a triangular graph, also called a ternary or trilinear diagram (see figure 3.5). When a system is optimized, it is important to be alert for

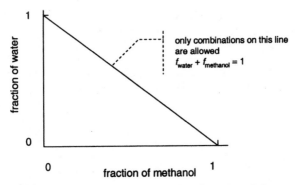

**Figure 3.4.** Optimizing a water–methanol mobile phase.

possible correlations among factors because this may influence the choice of method or the treatment of results.

### 3.2.4   Methods of Optimization

With due deference to the myriad mathematics dissertations and journal articles on the subject of optimization, I will briefly mention some of the general approaches to finding an optimum and then describe the recommended methods of experimental design in some detail. There are two broad classes that define the options: systems that are sufficiently described by a priori mathematical equations, called "models," and systems that are not explicitly described, called "model free." Once the parameters of a model are known, it is often quite trivial, via the miracles of differentiation, to find the maximum (maxima).

#### 3.2.4.1   Experimental Design

Experimental design lies between a fully characterized system and a system without a model. In this case it is accepted that a theoretical model of the system is not known but the response is sufficiently well behaved to be fitted to a simple linear or quadratic relationship to the factors (i.e., a straight or slightly curved line). Although I use the language of slopes and intercepts, do not confuse experimental design with calibration, in which there is often a known (linear) relationship between the indication of the measuring instrument and the concentration of analyte. The models of experimental design are used more in hope than certainty, but they do lead to improvement of the process. Take the example of optimizing the number of theoretical plates of a high-performance liquid chromatography (HPLC) experiment by changing the water/methanol ratio ($R$) and the con-

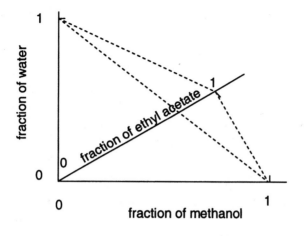

(a)

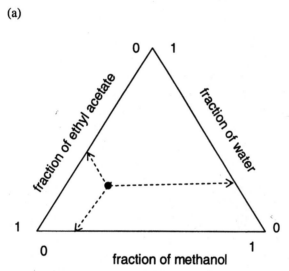

(b)

**Figure 3.5.** A three-component (mixture) system
whose values must total 1. (a) The allowed values
in three-dimensional space. (b) A ternary diagram
of the three-component system.

centration of acetic acid modifier ($A$). In experimental design the response is fitted to models such as:

$$N = \alpha + \beta_1 R + \beta_2 A + \beta_{12}(R \times A) \tag{3.1}$$

$$N = \alpha + \beta_1 R + \beta_2 A + \beta_{11} R^2 + \beta_{22} A^2 + \beta_{12}(R \times A) \tag{3.2}$$

How and why the response is fitted to these models is discussed later in this chapter. Note here that the coefficients $\beta$ represent how much the particular factor affects the response; the greater $\beta_1$, for example, the more $N$ changes as $R$ changes. A negative coefficient indicates that $N$ decreases as the factor increases, and a value of zero indicates that the factor has no effect on the response. Once the values of the factor coefficients are known, then, as with the properly modeled systems, mathematics can tell us the position of the optimum and give an estimate of the value of the response at this point without doing further experiments. Another aspect of experimental design is that, once the equation is chosen, an appropriate number of experiments is done to ascertain the values of the coefficients and the appropriateness of the model. This number of experiments should be determined in advance, so the method developer can plan his or her work.

An optimum is a maximum (or minimum), and this is described mathematically by a quadratic equation. The linear function of equation 3.1 can only go up or down, so the optimum will be at one end or the other of the factors. The absence of a maximum or minimum is often found in chemistry. Increasing temperature will speed reactions up, with the cutoff not being dependent on the temperature effect, which is always positive, but on the stability of compounds or other experimental factors. If there is a genuine maximum in the factor space, then this must be modeled by a quadratic (or greater power) equation. However, merely having fitted some data to an equation like equation 3.2 does not guarantee an optimum. The function might not have its maximum within the factor space. Perhaps the function happens to have a saddle point and does not go through a maximum or minimum at all. So your should not necessarily expect to find the optimum so easily.

### 3.2.4.2 Simplex Optimization

Another class of methods is based on model-free, hill-climbing algorithms (Deming and Morgan 1973). After some initial experiments in a likely region of factor space, the results indicate values at which to perform the next experiment, which in turn leads to further values, and so on, until a response is achieved that fits some predefined requirement. Ponder the hill-climbing analogy. You are lost in the fog on the side of a hill and want to climb to the top in the hope of seeing above the mist. Totally disoriented, you stagger about until a structured approach occurs. The algorithm is as follows. Take one step forward, one step back, one to the right, and one to the left. Decide which step took you the farthest upward and move to that location. Repeat until you are at the top. Mathematically, this is refined and embodied in a procedure known as the Simplex algorithm, and it is a remarkably powerful method of reaching the optimum. Rather than all the points of the compass, a Simplex is an $n + 1$ dimensional figure in the $n$-dimensional factor space (figure 3.6). The hill-climbing problem only requires a series of triangles (rather than the rectangles suggested in my simple solution offered above) to be mapped, with the worst-fitting point of a triangle being reflected in the line joining the other two (see figure 3.6). Starting with the triangle (Simplex) with vertices labeled 1, 2, and 3, the point 2 has the lowest response and so is reflected in the line joining 1 and 3 giving the new point 4. The new point plus the remaining two of the original triangle becomes the new Simplex. In figure 3.6, the new Simplex 1,3,4 becomes 3,4,5 when vertex 1 is reflected, and so on.

Simplex has been used in analytical method development. Its advantages are that the response should improve with each round of experiments, allowing the experimenter to decide when to discontinue the experiments; there are no arbitrary relations involved in the choice of the model equation; and the methodology used to select the next point can easily be implemented in a spreadsheet. Disadvantages are that the Simplex method is an open-ended procedure, for which the number of experiments depends on

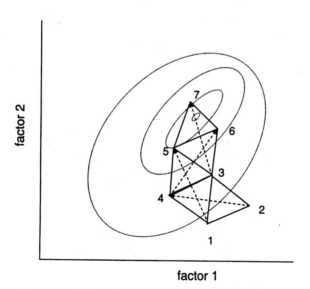

**Figure 3.6.** A Simplex optimization of a two-factor system. Numbers give the order of experiments. The first three experiments (points 1, 2, and 3) define the initial Simplex, and dashed-line arrows indicate which points are dropped in favor of new factor values.

the step size and the starting point, and it is prone to finding the nearest hill, not necessarily the highest hill (i.e., a local rather than a global optimum). Simplex is an optimization procedure, so the factors must already be chosen. The inclusion of a factor that has little or no effect can result in a lot of wasted effort.

### 3.2.4.3 Change One Factor at a Time

Ask anyone how they would investigate the effect of pH and temperature on the results of an experiment, and they would probably suggest an experiment that varied the temperature at constant pH, followed by more experiments at the now-fixed optimum temperature, this time changing the pH. Most would be astounded to learn that the final temperature–pH combination is not necessarily the optimum and that they have performed more experiments than they needed to. The problem lies in any correlation between the factors. When the effects of the factors are totally independent—that is, the changes in response with changes in temperature are the same whatever the pH, and vice versa—then change-one-factor-at-a-time approach does give the optimum within the step size of the factors (figure 3.7a). When there are significant interaction effects (e.g., nonzero values of $\beta_{12}$ in equations 3.1 and 3.2), then one pass through the change-one-factor-at-a-time approach does not achieve the optimum (figure 3.7b). The recommended methodology, experimental design, is discussed in more detail in the rest of this chapter.

## 3.3 Experimental Design

### 3.3.1 Which Factors Are Important?

Although the influences of some factors are self-evident, if experimental data are sparse, it can be difficult to determine all the significant factors. Experience will help provide a list of candidates, and the suite of factors to be investigated will be constrained by the time and effort available. At first, the linear coefficients of the model will give most information about the system. Equation 3.1 is written in a more general form for $k$ factors ($x_1, \ldots, x_k$) in equation 3.3. The linear coefficients comprise the "main effects," $\beta_i$, and the "two-way interaction effects" $\beta_{ij}$ The term $\varepsilon$ has been added to denote the error (i.e., the difference between the value of $Y$ [the response] calculated by the model and the true value). Epsilon is the combination of measurement error and lack of fit of the model.

$$Y = \alpha + \sum_{i=1}^{i=k} \beta_i x_i + \sum_{i=1}^{i=k} \sum_{j=i+1}^{j=k} \beta_{ij} x_i x_j + \text{higher order interaction} + \varepsilon \qquad (3.3)$$

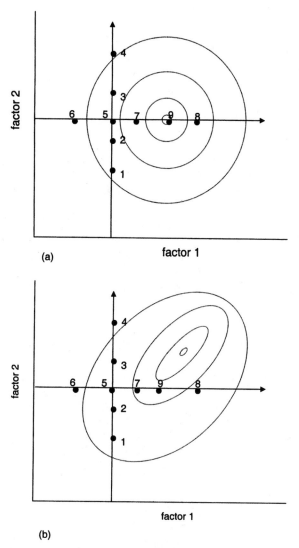

(a)

factor 1

(b)

factor 1

**Figure 3.7.** Why changing one factor at a time might not work. (a) A two-factor response (dotted circles) with independent effects. Possible experiments are shown, starting with keeping factor 1 constant and varying factor 2, and then keeping factor 2 constant at its optimum and varying factor 1. The result is near the true optimum. (b) A two-factor response (dotted ellipses) with correlated effects. A procedure similar to that shown in panel (a) no longer finds the optimum, although the final experiment gives the best result of the experiments performed.

When there are more than two factors, the possible interactions increase. With three factors there can be three, two-way interactions (1 with 2, 1 with 3, and 2 with 3), and now one three-way interaction (a term in $x_1 x_2 x_3$). The numbers and possible interactions build up like a binomial triangle (table 3.1). However, these higher order interactions are not likely to be significant, and the model can be made much simpler without loss of accuracy by ignoring them. The minimum number of experiments needed to establish the coefficients in equation 3.3 is the number of coefficients, and in this model there is a constant, $k$ main effects and $\frac{1}{2}k (k-1)$ two-way interaction effects.

Once factor values are determined, the minimum number of experiments is governed by the number of coefficients in the model, but if only the minimum number of experiments is done, there is no way to check the model. An analogy is drawing a straight line through two points, which cannot fail, but measurement of a third point will allow a least squares fit with some indication of the status of the linear model.

The starting point of an experimental design optimization is to survey as great a number of factors as is practicable and reasonable using a near-minimum set of experiments. Once the factors that are significant are identified, further experiments can be done to effect an optimization.

### 3.3.2  Randomization

It is important to note that, although the tables of experimental designs in this chapter show an orderly progression of experimental conditions, when the experiments are done, the order must be randomized. A random order can be generated in a spreadsheet, or if statistical or validation software is used, the program might randomize the order. Randomizing the order of experiments is the best way to confound uncontrolled effects and should always be employed.

Table 3.1. The numbers of coefficients in a linear effects model as a function of factors

| Factors ($k$) | Constant ($\alpha$) | Main effects ($\beta_i$) | Two-way interactions ($\beta_{ij}$) | Three-way interactions ($\beta_{ijm}$) | Four-way interactions ($\beta_{ijmn}$) |
|---|---|---|---|---|---|
| 1 | 1 | 1 | 0 | 0 | 0 |
| 2 | 1 | 2 | 1 | 0 | 0 |
| 3 | 1 | 3 | 3 | 1 | 0 |
| 4 | 1 | 4 | 6 | 4 | 1 |
| 5 | 1 | 5 | 10 | 10 | 5 |

## 3.4 Factorial Designs

Look at the experiments performed in figure 3.7. Even if the system is more like that depicted in figure 3.7a, where the nine experiments are near the optimum, in terms of determining the response of the experiment in factor space, the choice of values has kept to narrow ranges about the two selected fixed values of the factors. The area in the top right-hand corner is terra incognita as far as the experimenter is concerned. If nine experiments were done to find out about the response of a system, then perhaps the configurations of figure 3.8 would give more information.

Both of the patterns shown in figure 3.8 are common experimental designs and have a logic about them that recommends their use over the unstructured experiments of figure 3.7. The most straightforward pattern is one in which every combination of a chosen number of factor values is investigated. This is known as an $L$-level full-factorial design for $L$ values of each factor. For $k$ factors, $L^k$ experiments are indicated. A two-level design ($L = 2$) is most common for initial screening, and three-level designs are sometimes seen in optimization, but never more levels in a full-factorial design. With increasing $L$, the number of experiments becomes prohibitive, and there are better ways of gathering information. The two-level design is akin to drawing straight lines between two points. Such a design cannot detect curvature in a response, so it cannot be used to find an optimum (defined as a value greater than those around it), but it can be used to estimate main effects and interaction effects. Experimental points for two- and three-factor, two-level designs are shown in figure 3.9.

Figure 3.9 is another representation of the data in figure 3.2(a), with experiments performed at all combinations of the minimum and maximum values of each factor. The high and low values of a two-level experimental design are denoted + and −, respectively. The values +1 and −1 are "contrast coefficients" and are used when calculating the effects from experimental results. Remember that experimental designs can be done for any number of factors. The use of two and three factors in textbook examples is due to a problem of depicting higher dimensions. The number of experiments in a full factorial, two-level design is therefore 4, 8, 16, 32, ..., 1024 for 2, 3, 4, 5, ..., 10 factors, respectively, of which the two- and three-factor designs are shown in figure 3.9.

### 3.4.1  Calculating Factor Effects

Before the significance of the effect of a factor is considered, there must be an estimate of the effect. This is simply how much the response changes as the factor changes. It is a slope, and the value is a coefficient $\beta$ in equation 3.3. The task is therefore to find the coefficients of the model equation (e.g., equation 3.3). A positive $\beta_1$, for example, means that as the factor 1 increases, the response of the experiment also increases. A coefficient that is not

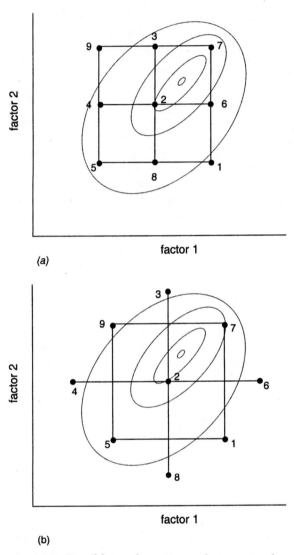

(a)

(b)

**Figure 3.8.** Possible configurations of nine sets of
factor values (experiments) that could be used to
discover information about the response of a
system. Dotted ellipses indicate the (unknown)
response surface. (a) A two-factor, three-level
factorial design; (b) a two-factor central composite
design.

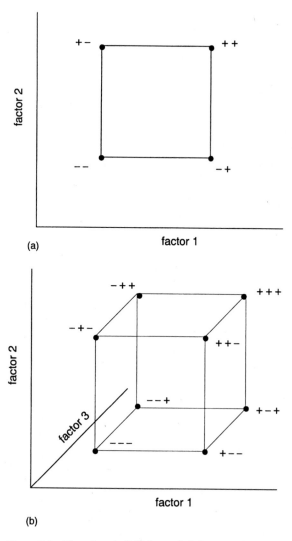

(a)

factor 2

factor 1

(b)

factor 2

factor 3

factor 1

**Figure 3.9.** Two-level, full factorial designs for (a) two and (b) three factors. The low value of each factor is designated by a contrast coefficient of— and the high value designated by +.

significantly different from zero means that the factor has no effect at all on the response. Calculating effects is easy if you know the contrast coefficients of the experimental design and, of course, the responses. Consider the table of contrast coefficients for experiments corresponding to the two-level, three-factor design shown in table 3.2.

Factor A might be temperature with a low of 20°C and high of 30°C. Factor B might be pH with a low level (–) of 5.5 and high (+) of 8.5, and C might be the HPLC column used with low a Dionex column and high a Waters column. Where the factor is an entity with discrete choices, the factor effect is simply the difference in response when the two entities are used. However, its significance can still be tested, even if optimization simply means choosing one or the other. I explain more about the choice of factor values later in this chapter. The full experimental design model for this three-factor system gives the response of experiment $i$, $Y_i$, as

$$Y_i = \alpha + \beta_A x_A + \beta_B x_B + \beta_C x_C + \beta_{AB} x_A x_B + \beta_{BC} x_B x_C$$
$$+ \beta_{CA} x_B x_A + \beta_{ABC} x_A x_B x_C + \varepsilon \quad (3.4)$$

where $x_A$ is the value of factor A, and so on. To generalize and aid interpretation, instead of computing values of $\beta$ that are the coefficients multiplying the values of the factors (and therefore having units of factor$^{-1}$), equation 3.4 is usually written

$$Y_i = \overline{Y} + b_A a + b_B b + b_C c + b_{AB} ab + b_{BC} bc + b_{CA} ca + b_{ABC} abc + \varepsilon \quad (3.5)$$

where $a$ is the contrast coefficient (+1 or -1) for A, and $\overline{Y}$ is the average response of all the experiments and is also the response for factors with con-

Table 3.2. Contrast coefficients for a two-level, three-factor, full-factorial experimental design

| Experiment | Factor | | |
|---|---|---|---|
| | A | B | C |
| 1 | – | – | – |
| 2 | + | – | – |
| 3 | – | + | – |
| 4 | + | + | – |
| 5 | – | – | + |
| 6 | + | – | + |
| 7 | – | + | + |
| 8 | + | + | + |

Minus (–) represents the low level of the factor, and plus (+) the high level. The order in which the experiments are performed should be randomized.

trast coefficients of zero. The coefficients $b$ have the dimension of $Y$ and are the changes in the response as the factors go from low values to high values.

For this particular design, there are only as many coefficients as experiments (8), and so there no way to estimate how good the model is ($\varepsilon$). As the number of factors increases, the possible experiments that can be done ($N = L^k$) increases faster than the number of coefficients in the model (see fractional designs below).

The symmetry of the contrast coefficients is quite beautiful. When constructing designs, it is easy to see if a mistake has been made because breaking the symmetry is readily apparent. The experiments (runs) can be grouped in several ways. Consider runs (1,2), (3,4), (5,6), and (7,8). Each pair is characterized by factor A going from low to high, while factors B and C do not change. Therefore, if the main effect of factor A is of interest, it should be estimated by the average of the differences in responses of these experiments. To say this another way, if only A is changing in these pairs, then it alone must be responsible for the change in response. Therefore

$$b_A = \frac{(Y_2 - Y_1) + (Y_4 - Y_3) + (Y_6 - Y_5) + (Y_8 - Y_7)}{4} \tag{3.6}$$

where $Y_i$ is the response of the $i$th experiment. It is possible to pick out four pairs of experiments for each factor that have the structure of keeping the other two factors constant and allowing that factor to change from low to high. These four pairs are

$$b_B = \frac{(Y_3 - Y_1) + (Y_4 - Y_2) + (Y_7 - Y_5) + (Y_8 - Y_6)}{4} \tag{3.7}$$

$$b_C = \frac{(Y_5 - Y_1) + (Y_6 - Y_2) + (Y_7 - Y_3) + (Y_8 - Y_4)}{4} \tag{3.8}$$

Here is the power of experimental design. From only 8 experiments, 3 parameters have been estimated 12 times. Further scrutiny of table 3.2 shows pairs of experiments that will allow estimation of interaction effects. What is the AB interaction effect? Mathematically in the model of equation 3.5, the AB interaction effect is the coefficient $b_{AB}$, the effect of A on the effect of B or, equivalently, the effect of B on the effect of A. The implication is that there is a different change when A goes from low to high, when B is low, than when B is high. The value of the interaction effect is the change of the change. This is illustrated more clearly in figure 3.10.

Like main effects, interaction effects can be positive, negative, or zero. Higher order effects are viewed in the same way: The three-way interaction effect ABC is the change in the change of the change of response as A goes

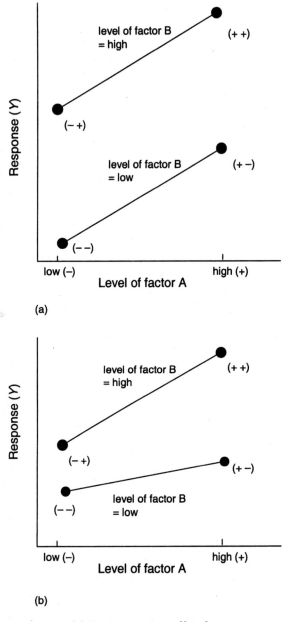

**Figure 3.10.** (a) No interaction effect between factors A and B. The change in response as factor A goes from low to high is the same whether B is low or high. (b) Positive interaction effect between factors A and B. The change in response as factor A goes from low to high is greater when factor B is high than when it is low.

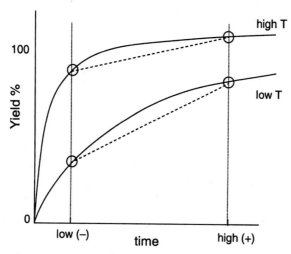

**Figure 3.11.** Example of a negative interaction effect between temperature and time on the yield of a synthetic product.

from low to high, as B goes from low to high as C goes from low to high. It may be appreciated why high-order effects are usually insignificant, as the existence of such multi-way effects is difficult to conceive. Two-way effects are not uncommon, however. For example, if the yield of a synthesis (this is the response) is being studied as a function of time and temperature, an interaction effect might be expected. At low temperatures the effect of increasing time is likely to be greater than at high temperatures, when the reaction might be completed quickly. In figure 3.11, the increase of yield is less pronounced at high temperatures because the reaction is kinetically activated, and therefore the rate increases with temperature.

Note that the approximations of straight lines imposed on the responses (in figure 3.11 the dotted lines joining the experimental points represent the experimental design model) certainly do not follow the real profiles of the experiment, but the results would be good enough to give some idea of the effects and indicate that there was, indeed, an interaction. Consider Table 3.2: Runs 1 and 2 and 5 and 6 are when A goes from high to low while B is low. C doesn't matter as long as it is the same. Runs 3 and 4 and 7 and 8, are when A goes from high to low when B is high. The interaction effect is the difference between these averages divided by the number of differences (here 4).

$$b_{AB} = \frac{\left[(Y_2 - Y_1) + (Y_6 - Y_5)\right] - \left[(Y_4 - Y_3) + (Y_8 - Y_7)\right]}{4} \qquad (3.9)$$

There is a very easy way to perform the calculations. Multiply the response (Y) by the contrast coefficient of the effect to be estimated, sum the values,

and divide by half the number of runs. For interaction effects, multiply the −
and + values to give an interaction column. For example, the contrast coefficients for the interaction between factors A and B (usually written AB) are
calculated by multiplying the contrast coefficients of A by the equivalent
coefficients of B. Table 3.3 gives the contrast coefficients for all two-way interactions and the three-way interaction for the two-level, three-factor design
of table 3.2. According to this formulation, the main effect for A is

$$b_A = \frac{-Y_1 + Y_2 - Y_3 + Y_4 - Y_5 + Y_6 - Y_7 + Y_8}{4} \tag{3.10}$$

and the three-way interaction ABC is

$$b_{ABC} = \frac{-Y_1 + Y_2 + Y_3 - Y_4 + Y_5 - Y_6 - Y_7 + Y_8}{4} \tag{3.11}$$

### 3.4.2  Uncertainty of an Effect Estimate

As I have shown, the response given by the model equation (3.5) has an error
term that includes the lack of fit of the model and dispersion due to the measurement (repeatability). For the three-factor example discussed above, there
are four estimates of each effect, and in general the number of estimates are
equal to half the number of runs. The variance of these estimated effects gives
some indication of how well the model and the measurement bear up when
experiments are actually done, if this value can be compared with an expected
variance due to measurement alone. There are two ways to estimate measurement repeatability. First, if there are repeated measurements, then the standard deviation of these replicates ($s$) is an estimate of the repeatability. For
$N/2$ estimates of the factor effect, the standard deviation of the effect is

Table 3.3.  Contrast coefficients for interaction effects of a two-level, three-
factor, full-factorial experimental design

| Experiment | Factor | | | | | | |
|---|---|---|---|---|---|---|---|
| | A | B | C | AB | BC | CA | ABC |
| 1 | − | − | − | + | + | + | − |
| 2 | + | − | − | − | + | − | + |
| 3 | − | + | − | − | − | + | + |
| 4 | + | + | − | + | − | − | − |
| 5 | − | − | + | + | − | − | + |
| 6 | + | − | + | − | − | + | − |
| 7 | − | + | + | − | + | − | − |
| 8 | + | + | + | + | + | + | + |

$$s_{effect} = \sqrt{\frac{2s^2}{N}} \qquad (3.12)$$

In the case where duplicate measurements are made and there is no other estimate of the measurement precision, the standard deviation of the effect is calculated from the differences of the pairs of measurements ($d_i$)

$$s_{effect} = \sqrt{\frac{\sum d^2}{2N}} \qquad (3.13)$$

Second, if each run is performed only once, the effect standard deviation can still be estimated because high-order effects should be zero. A non-zero estimate of a third-order effect, therefore, may be attributed to random error and used to estimate the standard deviation of all effects. If $m$ high-order effects can be calculated, the standard deviation of the effect is estimated as

$$s_{effect} = \sqrt{\frac{\sum_{i=1}^{i=m} E_i^2}{m}} \qquad (3.14)$$

If there is only one high-order effect ($E$), then equation 3.14 is simply $s_{effect} = E$.

How is this used? If an effect were really zero, then estimates of that effect should have a mean of zero and standard deviation $s_{effect}$. Therefore a $t$ test done at $N/2$ degrees of freedom can give the probability of the null hypothesis that the factor effect is zero from a calculation

$$t = \frac{|E - 0|}{s_{effect}} \qquad (3.15)$$

and the probability $\alpha = Pr(T \geq t)$. If $\alpha$ is small (say, $< 0.05$), the null hypothesis is rejected at the $100(1 - \alpha)\%$ (e.g., 95%) level, and we can conclude that the effect is indeed significant. The probability is calculated in Excel by =TDIST(t, N/2, 2). Alternatively, if 95% is a given decision point, the 95% confidence interval on the effect is $E \pm t_{0.05",N/2} \times s_{effect}$, and if this encompasses zero, then we can conclude that the value of the effect cannot be distinguished from zero. The Student's $t$ value is computed in Excel by =TINV(0.05,N/2).

### 3.4.3  Fractional Designs

It is not unusual to want to investigate 10 factors, but it is unusual, and not necessary, to do 1024 experiments to discover 10 main effects and 45 two-way effects. However, care must be taken when deciding not to perform experiments at all possible combinations. Consider the case of three factors (figure 3.9b). Suppose only four experiments were possible to estimate the constant, plus the three main effects. There are four coefficients and four experiments, so just enough for the math, but which four? Does it matter? Yes, there are only two combinations of four experiments that allow calculation of all main effects, and these are shown in figure 3.12.

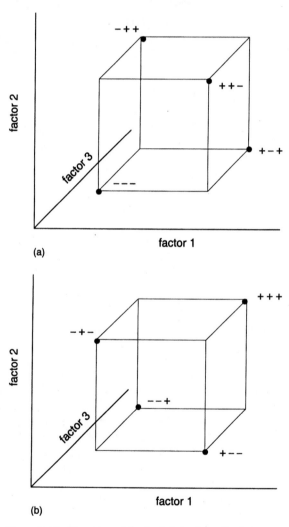

**Figure 3.12.** Fractional factorial designs for two levels and two factors that can be used to estimate the main effects and constant. The experiments in (a) and (b) are complements of each other. Either design can be used.

The problem with just leaving out four experiments at random is that in the remaining combinations of factor levels, there must be at least one high and one low for each factor. The two sets of four combinations shown in figure 3.12 are the only way to do this.

In general, a factorial design is fractionated by deciding which high-order interaction effects are to be "aliased" or "confounded" with the main effects. Doing fewer experiments means that the ability to discover every possible high-order interaction effect is sacrificed, and these effects are lumped in with the estimates of main and lower order interaction effects. If one assumes that interactions of order greater than two are insignificant, then it does not matter if the interactions are added to main effects, as adding zero does not change the estimate of the main effect. The choice is made by manipulating contrast coefficients. If a particular $j$-way interaction is to be aliased with a main effect (the main effect factor is not included in the $j$ factors), then the experiments for which the product of the contrast coefficients of the $j$ effects is the same as that of the main effect should be performed. The rules of multiplying signs mean that, for example, $+ - +$ and $- - -$ give an overall $-$, while $- - +$ and $+ + +$ give $+$. If the procedure goes well, half the experiments will have been chosen, and if desired and it is possible, they can be repeated for a different set of aliases. A half design in which the three-way effect of factors 1, 2, and 3 is aliased with the main effect of factor 4 in a two-level, four-factor design is given in table 3.4.

Table 3.4.  Fractionating a two-way, four-factor design by aliasing the three-way $1 \times 2 \times 3$ effect with the main effect of 4.

| Experiment | 1 | 2 | 3 | 4 | $1 \times 2 \times 3$ |
|---|---|---|---|---|---|
| 1[a] | + | + | + | + | + |
| 2 | − | + | + | + | − |
| 3 | + | − | + | + | − |
| 4[a] | − | − | + | + | + |
| 5 | + | + | − | + | − |
| 6[a] | − | + | − | + | + |
| 7[a] | + | − | − | + | + |
| 8 | − | − | − | + | − |
| 9 | + | + | + | − | + |
| 10[a] | − | + | + | − | − |
| 11[a] | + | − | + | − | − |
| 12 | − | − | + | − | + |
| 13[a] | + | + | − | − | − |
| 14 | − | + | − | − | + |
| 15 | + | − | − | − | + |
| 16[a] | − | − | − | − | − |

[a] $1 \times 2 \times 3$ effect = effect of 4.

All fractional factorial designs give numbers of experiments in which the exponent of the number of factors is reduced. A full factorial design at $L$ levels on $k$ factors requires $L^k$ experiments. When fractionated this is reduced to $L^{k-1}$, or $L^{k-2}$, and so on; in general to $L^{k-p}$, where $p$ is an integer less than $k$. In the example in table 3.4 the design is halved from $2^4 = 16$ to $2^{4-1} = 8$, and in the simple three-way design from $2^3 = 8$ to $2^{3-1} = 4$. The resolution of a fractional design ($R$) is defined as the property that no $m$-way effect is aliased with an effect greater than $R - m$. Thus, a design of resolution III (resolution is usually given as a Roman numeral) means that main effects ($m = 1$) are aliased with $3 - 1 = 2$–way effects, but not other main effects. This is the case for both examples given here. A resolution IV design means that main effects are aliased with three-way effects ($4 - 1$), but two-way effects can be aliased with each other ($4 - 2 = 2$). Experimental design software usually has a facility to calculate fractional designs of a particular resolution.

## 3.5 Plackett–Burman Designs

Method validation (chapter 8) involves the effects of factors on the results of experiments. A method is robust if ordinary changes, such as small fluctuations in temperature or changing a source of reagent, do not significantly affect the measurement result. To claim that a method is robust in respect of certain parameters needs only the main effects of those parameters to be shown to be negligible. A highly fractionated design, which is ideal for method validation, was first reported by Plackett and Burman (1946). In their design a minimum $4n$ experiments is required to estimate the main effects of $4n - 1$ factors. If there are 3, 7, or 11 factors, then 4, 8, or 12 experiments, and so on, are performed to obtain estimates of the main effects. If there are, say, only six factors of interest, the extra factor can be used as a dummy factor to assess the model error. In a dummy experiment a parameter is chosen that does not influence the result. I prefer testing the influence of my rotation on the result: In the − setting I turn once clockwise before making the measurement, and in the + setting I turn once counterclockwise. The gyrations of the analyst do not make any real difference to the ultimate analytical result, so the value of the dummy main effect should be zero, within experimental error, and can be used to estimate the expected random component of the result. The Plackett and Burman design works only for multiples of 4 experiments, so if 8 factors have been identified, then 12 experiments must be done for 11 factors. Either three dummy factors are added, or some other real factors to be investigated can be found.

Each experiment is performed at one of two levels of each factor, just as for the two-level factorial designs described above. The contrast coefficients are a consequence of by the method and are generated from the conditions of a seed experiment by advancing each designated level around the factors. Table 3.5 gives the contrast coefficients for experiments that must be carried out for a seven-factor Plackett-Burman design.

In table 3.5, the factors A–G are the parameters being studied: temperature, mobile phase, pH, dummy, etc. Note that one experiment is done at all low levels (experiment 8). If the low (–) level is chosen as the nominal value for the experiment, the result of experiment 8 is the benchmark. Consider the structure of table 3.5. All the pluses and minuses in the first seven experiments are in a line diagonally and wrap over from factor G to factor A. In fact, after the first row, for experiment 1, each subsequent row is just the same order of signs moved over one factor, with the sign for factor G wrapping back to the next factor A. As there are seven factors, if the process were repeated for experiment 8, this would be a repeat of experiment 1, so the last experiment terminates the process with all minuses. Note that each factor (down a column) has four minus signs and four plus signs. The seed contrast coefficients for $n = 2$ to 5 are given in table 3.6. To use them, arrange the given contrast coefficients in the first row in a table, as shown in table 3.5. Then create the next $4n - 2$ rows by moving the coefficients over one factor and wrapping from the last factor to the first. The last row is all minuses.

The value of an effect and its standard deviation are calculated in the same way as for factorial designs. Multiply the responses by their contrast coefficients for a given factor, sum them, and divide this number by half the number of experiments ($2n$, for $4n$ experiments) to give the value of the effect. With no high-order interactions available in the design, either an independent estimate of repeatability or the use of dummy variables is essential. For $m$ dummy variables, the effect standard deviation is calculated using equation 3.14, where $E_i$ is the measured effect of the $i$th dummy factor. The significance of the effect is then determined by a Student's $t$ test described earlier.

Table 3.5. The designated levels as contrast coefficients for a seven-factor, eight-experiment Plackett-Burman design.

| Experiment | A | B | C | D | E | F | G |
|---|---|---|---|---|---|---|---|
| 1 | + | + | + | – | – | + | – |
| 2 | – | + | + | + | – | – | + |
| 3 | + | – | + | + | + | – | – |
| 4 | – | + | – | + | + | + | – |
| 5 | – | – | + | – | + | + | + |
| 6 | + | – | – | + | – | + | + |
| 7 | + | + | – | – | + | – | + |
| 8 | – | – | – | – | – | – | – |

Table 3.6. Seed contrast coefficients for Plackett-Burman experimental designs.

| $n^a$ | Contrast coefficients for the first experiment |
|---|---|
| 2 | + + + − − + − |
| 3 | + + − + + + − − − + − |
| 4 | + + + + − − − − − − + + + − + |
| 5 | + − + + − − − − + − + − + + + + − − + |

[a]where the number of factors studied is $4n - 1$ in $4n$ experiments

## 3.6 Central Composite Designs

The central composite design is the workhorse design for determining an optimum. Each factor has experiments performed at one of five levels, thus allowing one to fit a quadratic model that has opportunity for turning points (maximum or minimum). It is not a five-level factorial design (the experiments become prohibitive very quickly; e.g., with $5^k$, $k = 2$ would be as far as anyone might want to go), but it takes a two-level design and adds a center point and star configuration of experiments that may go outside the two-level design (figure 3.13).

With five levels, contrast coefficients are not simply plus and minus, but give the spacing of the values. They are sometimes referred to as "coded" values. It is possible to choose designs where the star points are outside of the two-level design (central composite circumscribed), wholly inside the two-level design (central composite inscribed), or in the face of each side of the two-level design (central composite face centered). The first configuration is most often encountered, but if there are limits on factor values defined by the two-level design, then face centered or inscribed are used. If the corner and outer points can fit on a circle (sphere, or $n$-dimensional equivalent), the design is said to be rotatable. In this case the five levels of points are $-\alpha, -1, 0, +1, +\alpha$; where $\alpha = n_c^{1/4}$, and $n_c$ is the number of points in the cube. So for 2, 3, 4, and 5 factors, $\alpha = 1.41, 1.68, 2.00,$ and 2.38 respectively. Table 3.7 gives the 15 experiments that must be done for a three-factor, rotatable, central composite circumscribed design. As always, the order of the experiments is randomized, and replicates should be done at the center point. The values each factor takes are mapped on to the $-\alpha, -1, 0, +1, +\alpha$ scheme. For example, in the three-factor design, suppose one factor is temperature and you select 10° and 30°C as the −1 and +1 levels . The center point is half way between −1 and +1, and therefore is at 20°C. The lowest point at $-\alpha$ then comes in at $20 - (30 - 20) \times 1.682$ = 3.18°C and the highest $(+\alpha)$ at $20 + (30 - 20) \times 1.682 = 36.82°C$. This calculation starts with the center point (0), which is (−1 value) + [(+1 value) − (−1 value)]/2. The calculation [(+1 value) − (−1 value)]/2 gives the change

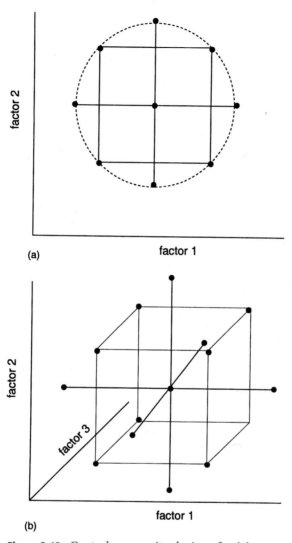

**Figure 3.13.** Central composite designs for (a) two and (b) three factors. The design in panel a is rotatable.

in factor value corresponding to a change in 1 in the coded values. Therefore a change in $\alpha$ from zero in the coded values gives a change in [(+1 value) − (−1 value)]/2 from zero in the real factor values. Hence -$\alpha$ is (0 value) − $\alpha$ × [(+1 value) − (−1 value)]/2, and +$\alpha$ is (0 value) + $\alpha$ × [(+1 value) − (−1 value)]/2. These calculations are easier with actual numbers, and most experimental design software can calculate these levels for the user.

Table 3.7. Levels for a three-factor, rotatable, central composite design.

| Experiment | Factor | | |
|---|---|---|---|
| | A | B | C |
| 1[a] | −1 | −1 | −1 |
| 2[a] | +1 | −1 | −1 |
| 3[a] | −1 | +1 | −1 |
| 4[a] | +1 | +1 | −1 |
| 5[a] | −1 | −1 | +1 |
| 6[a] | +1 | −1 | +1 |
| 7[a] | −1 | +1 | +1 |
| 8[a] | +1 | +1 | +1 |
| 9[b] | −1.682 | 0 | 0 |
| 10[b] | +1.682 | 0 | 0 |
| 11[b] | 0 | −1.682 | 0 |
| 12[b] | 0 | +1.682 | 0 |
| 13[b] | 0 | 0 | −1.682 |
| 14[b] | 0 | 0 | +1.682 |
| 15[c] | 0 | 0 | 0 |

[a]Full two-level factorial design. [b]Star design. [c]Center point.

### 3.6.1  Models and Factor Effects

Typical models for two- and three-factor systems subjected to a central composite design are

$$Yi = \bar{Y} + b_A a + b_B b + b_{AA} a^2 + b_{BB} b^2 + b_{AB} ab + \varepsilon \qquad (3.16)$$

$$Y_i = \bar{Y} + b_A a + b_B b + b_C c + b_{AA} a^2 + b_{BB} b^2 + b_{CC} c^2$$
$$+ b_{AB} ab + b_{BC} bc + b_{CA} ca + \varepsilon \qquad (3.17)$$

Dedicated software is often used to fit the data, but it is also possible to calculate the coefficients easily in Excel. Equations 3.16 and 3.17 are linear in the coefficients of the equation and therefore can be obtained by the usual methods such as LINEST. Create columns for the values of each of the factor codes, then add extra columns for the square of these codes ($a^2$, etc.) and their products ($ab$, etc.). When fitting these with LINEST, chose the whole matrix for the $x$ values and the column containing the responses for $y$. The LINEST output array is now $p$ columns by five rows, where $p$ is the number of coefficients, including the mean response.

### 3.6.2  Factor Standard Deviations

You should always replicate the center point; the standard deviation of the effect is calculated from the standard deviation of these replicates. The significance of any effects is tested as before, with $n - p$ degrees of freedom ($n$ is the number of experiments and $p$ the number of coefficients calculated).

## 3.7 Other Designs

There are a number of other designs that have advantages in particular circumstances. Many are just variants on the theme of maximally spanning the factor space. The Box-Behnken design is a two-level, spherical, rotatable design (Box and Behnken 1960). For three factors it has experiments at the center and middle of each edge of a cube (figure 3.14).

The uniform shell or Doehlert design (Doehlert 1970) has fewer points than the central composite design, but it still fills the factor space. It is derived from a simplex, which is the uniform geometrical shape having $k + 1$ vertices for $k$ factors. Thus, for two factors the simplex is an equilateral triangle, and for three factors it is a regular tetrahedron. The rest of the design is obtained by creating points by subtracting every point from every other point. To illustrate this for two factors, start with a triangle with coordinates (0,0), (1,0), (0.5, 0.866). Subtraction generates further points at (−1,0), (−0.5, −0.866), (−0.5, +0.866) and (0.5, −0.866). These describe a hexagon with a center point (see figure 3.15).

D-optimal designs (Mitchell 1974) are used when a subset of candidate experimental points are chosen to optimize some criterion, such as the greatest volume enclosed by the points or the maximum distance between all points. These designs are very useful under two circumstances. The first is when a traditional design cannot be used because of experimental reasons. In an optimization of temperature and concentration, the combination of lowest temperature and highest concentration may be ruled out because of low solubility. With any constraints that limit the factor space, a traditional design with smaller spacing is mapped over the space, including overlapping the areas without data. A smaller subset is chosen to optimize D, and these become the experiments to be performed. The second circumstance arises when the factor to be optimized can only take a limited number of values that are set by the nature of the factor. In experiments that require solutes of different solubility, the experimenter can choose from the candidate set, but cannot change the solubilities. A D-optimal design then takes the values that best fit the problem. Sometimes the choice of the design points is not easy. The solution to the smallest total distance is akin to the traveling salesman problem, in which a salesman must visit towns via the shortest total route, and it is an "NP-complete" problem. There is no algorithmic solution, and the number of possible solutions scales as a power law in the number of points (in the salesman example, towns). However, there are approximate solutions, and searching algorithms such as genetic algorithms can also yield solutions.

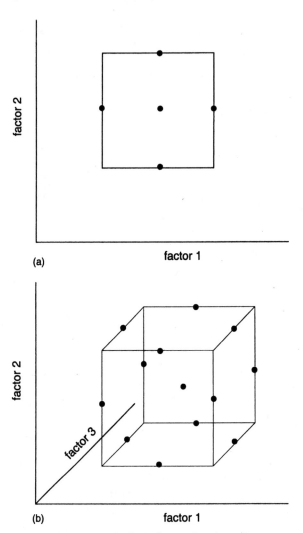

**Figure 3.14.** Box-Behnken design for three factors. The design is spherical and rotatable.

## 3.8 Worked Example

Here is an example from my research group's work on biosensors. The sensor has an electrode modified with a peptide that binds a target metal ion, which is then electrochemically reduced, the current being proportional to the concentration of metal. The electrode is calibrated with solutions of known concentration. Two experimental designs were used. The first was a two-level factorial design to find the effects of temperature, pH, added salt, and accumulation time. Spreadsheet 3.1 shows the design.

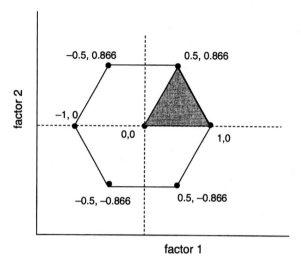

factor 2

factor 1

**Figure 3.15.** Doehlert design for two factors, with coded values. The shaded triangle is the generating simplex.

The order of experiments was randomized : 2, 6, 12, 8, 16, 13, 10, 15, 14, 11, 4, 1, 7, 5, 3, 9. Each experiment was performed in duplicate, and a pooled standard deviation was calculated as 4.14 µA cm$^{-2}$. From equation 3.12, with $N = 16$, $s_{effect} = \sqrt{(4.14^2 / 8)} = 1.46$ µA cm$^{-2}$. The main effects were calculated by summing the contrast coefficient for the effect and experiment multiplied by the response (current) from each experiment, dividing by half the number of experiments (here 8). The significance of each effect was tested by computing the $t$ value of the effect divided by $s_{effect}$. In each case the probability of finding an equal or greater $t$ value in repeated experiments on the system assuming the effect is not significant is less than 0.000 (i.e., all the effects are significant at least at the 99.9% level). Note that the effect of added salt has a negative effect; that is, adding 500 mmol L$^{-1}$ sodium chloride reduces the current by an average of 9.48 µA cm$^{-2}$.

The second experimental design was for an interference study performed with a Plackett-Burman design. A cadmium electrode was run in a solution of 0.2 µM Cd$^{2+}$ to which a low level (0.05 µM) or high level (0.5 µM) of potentially interfering ions Pb$^{2+}$, Cu$^{2+}$, Ni$^{2+}$, Zn$^{2+}$, Cr$^{3+}$, Ba$^{2+}$ had been added. Recall that in a Plackett-Burman design there are $4n$ experiments for $4n - 1$ factors, so one dummy factor was added to the list to make 7 factors total. The response was the current, and the aim was to determine whether any of the metals caused a significant change in the current when they went from low to high concentration. The design shown in spreadsheet 3.2 was used, after suitably randomizing the order. The results (currents) are shown in column J, returned to the logical order in which the contrast coefficients were generated.

|  | A | B Temperature | C pH | D salt | E time | F Response | G s | R |
|---|---|---|---|---|---|---|---|---|
| 10 |  | Temperature | pH | salt | time | Response | s | R |
| 11 | 1 | -1 | -1 | -1 | -1 | 2.30 | 0.37 |  |
| 12 | 2 | 1 | -1 | -1 | -1 | 9.15 | 2.69 |  |
| 13 | 3 | -1 | 1 | -1 | -1 | 9.79 | 3.63 |  |
| 14 | 4 | 1 | 1 | -1 | -1 | 19.84 | 3.91 |  |
| 15 | 5 | -1 | -1 | 1 | -1 | 1.96 | 0.46 |  |
| 16 | 6 | 1 | -1 | 1 | -1 | 4.76 | 2.40 |  |
| 17 | 7 | -1 | 1 | 1 | -1 | 9.31 | 2.08 |  |
| 18 | 8 | 1 | 1 | 1 | -1 | 17.59 | 1.47 |  |
| 19 | 9 | -1 | -1 | -1 | 1 | 6.72 | 1.50 |  |
| 20 | 10 | 1 | -1 | -1 | 1 | 32.32 | 7.57 |  |
| 21 | 11 | -1 | 1 | -1 | 1 | 34.34 | 3.55 |  |
| 22 | 12 | 1 | 1 | -1 | 1 | 60.49 | 9.55 |  |
| 23 | 13 | -1 | -1 | 1 | 1 | 1.09 | 0.68 |  |
| 24 | 14 | 1 | -1 | 1 | 1 | 25.11 | 6.93 |  |
| 25 | 15 | -1 | 1 | 1 | 1 | 15.44 | 0.69 |  |
| 26 | 16 | 1 | 1 | 1 | 1 | 23.82 | 3.62 |  |
| 27 | Effect | 14.02 | 13.40 | -9.48 | 15.58 | 17.13 |  |  |
| 28 | t = effect / $S_{effect}$ | 9.58 | 9.16 | -6.49 | 10.65 | $S^2_{pool}$ | 17.10603 |  |
| 29 | p = TDIST(t,n/2,2) | 0.0000 | 0.0000 | 0.0002 | 0.0000 | $S^2_{effect}$ | 2.138254 |  |

Formula annotations:

=SUMPRODUCT(B11:B26,$F$11:$F$26)/8

=B27/SQRT(G29)

=TDIST(B28,8,2)

=SUMSQ(G11:G26)/16

=G28/8

=AVERAGE(F11:F26)

**Spreadsheet 3.1.** Results of a two-level full factorial experimental design to investigate the effects of temperature (-1 = 10°C, +1 = 35°C), pH (-1 = 3, +1 = 9), added salt (-1 = no added salt, +1 = 500 mmol L$^{-1}$), and accumulation time (-1 = 1 min, +1 = 15 min) on the current of a modified electrode detecting Cu$^{2+}$.

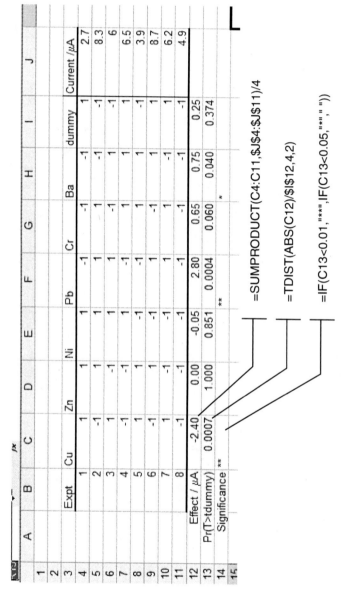

| | Expt | Cu | Zn | Ni | Pb | Cr | Ba | dummy | Current /μA |
|---|---|---|---|---|---|---|---|---|---|
| | 1 | 1 | 1 | 1 | -1 | -1 | -1 | 1 | 2.7 |
| | 2 | -1 | 1 | 1 | 1 | 1 | -1 | -1 | 8.3 |
| | 3 | 1 | -1 | 1 | 1 | -1 | 1 | -1 | 6 |
| | 4 | -1 | -1 | 1 | -1 | 1 | 1 | 1 | 6.5 |
| | 5 | 1 | 1 | -1 | -1 | 1 | 1 | -1 | 3.9 |
| | 6 | -1 | 1 | -1 | 1 | -1 | 1 | 1 | 8.7 |
| | 7 | 1 | -1 | -1 | 1 | 1 | -1 | 1 | 6.2 |
| | 8 | -1 | -1 | -1 | -1 | -1 | -1 | -1 | 4.9 |
| Effect / μA | | -2.40 | 0.00 | -0.05 | 2.80 | 0.65 | 0.75 | 0.25 | |
| Pr(T>tdummy) | | 0.0007 | 1.000 | 0.851 | 0.0004 | 0.060 | 0.040 | 0.374 | |
| Significance | ** | ** | | | ** | | * | | |

=SUMPRODUCT(C4:C11,$J$4:$J$11)/4

=TDIST(ABS(C12)/$J$12,4,2)

=IF(C13<0.01,"**",IF(C13<0.05,"*"," "))

**Spreadsheet 3.2.** Results of a Plackett-Burman design to discover if interfering metals have a significant effect on the current of a peptide biosensor for cadmium.

The effects are calculated easily by the Excel function SUMPRODUCT—for example, the effect of copper is =SUMPRODUCT(C2:C9,$J2:$J9)/4. This can be copied across the row to calculate the rest of the effects.[2] At first sight it looks as if copper had a possibly significant negative effect and lead a positive one. The rest, including the dummy, are less convincing. The significance of the effects were tested in two ways, using a Rankit plot and comparing the effects to the dummy. In a Rankit plot, the distribution of the results are compared to the expected normal distribution. A straight line through zero indicates a normal distribution of the effects, whereas outliers at the ends of the line point to significant, non-zero effects that add to (or subtract from) the underlying random distribution (see also chapter 2).

The calculation for the Rankit plot is shown in spreadsheet 3.3. The effects are ordered from most negative to most positive. A column of the rank of each effect is then created, with ties (none here) taking the higher rank (e.g., 1, 2, 4, 4, 5). The column headed $z$ is the point on the cumulative normal distribution of the rank/$(N + 1)$, where $N$ is the number of experiments. The $z$ score is calculated by the function =NORMSINV(z). When this is plotted against the effect (see figure 3.16), it is clear that copper and lead do, indeed, appear to be off the line, and all the other effects are concluded to be insignificant.

Alternatively, the dummy effect can be taken as the repeatability of the factor effects. Recall that a dummy experiment is one in which the factor is chosen to have no effect on the result (sing the first or second verse of the national anthem as the −1 and +1 levels), and so whatever estimate is made must be due to random effects in an experiment that is free of bias. Each factor effect is the mean of $N/2$ estimates (here 4), and so a Student's $t$ test can be performed of each estimated factor effect against a null hypothesis of the population mean = 0, with standard deviation the dummy effect. Therefore the $t$ value of the $i$th effect is:

| | A | B | C | D | |
|---|---|---|---|---|---|
| 27 | | | | | |
| 28 | Interferent | Effect | Rank | z | =NORMSINV(C29/8) |
| 29 | Cu | -2.4 | 1 | -1.15035 | |
| 30 | Ni | -0.05 | 2 | -0.67449 | |
| 31 | Zn | 0 | 3 | -0.31864 | |
| 32 | dummy | 0.25 | 4 | 5.47E-10 | |
| 33 | Cr | 0.65 | 5 | 0.31864 | |
| 34 | Ba | 0.75 | 6 | 0.67449 | |
| 35 | Pb | 2.8 | 7 | 1.15035 | |
| 36 | | | | | |

**Spreadsheet 3.3.** Data for the Rankit plot of interference effects in a cadmium biosensor determined from a Plackett-Burman experimental design. See figure 3.16 for the plot.

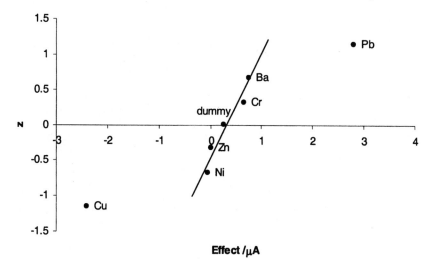

**Figure 3.16.** Rankit plot of the effects of interferences on a cadmium biosensor, determined from a Plackett-Burman design. See spreadsheet 3.3 for data.

$$t_i = \frac{|E_i - 0|}{E_{dummy}}$$

(3.18)

The probabilities of each effect are given in spreadsheet 3.2, calculated by =TDIST(t, df, tails) in row 13. Note the flagging of significant effects using the =IF function. Often it is as easy to decide the significance of an effect manually, but if many calculations are to be done, the automation is worth it. This calculation confirms the significance of the copper and lead interferences, but it also suggests that barium may interfere. This is the problem of interpreting statistics with only a few data, and the analyst should always use proper judgment. There are good chemical reasons that copper and lead would interfere in the way they do, but barium was not thought to be a significant source of interference (i.e., we went with the Rankit plot rather than the *t* test). If the issue were important, then more experiments would have to be done to obtain a better estimate of the barium effect or to obtain a better estimate of the standard deviation of an effect.

## Notes

1. In the realms of chaos and catastrophe theory, there are systems that have different values depending on their past history, but chemical analysis may be considered to be (mostly) well behaved.
2. The dollar sign in front of the column letter ($J) is to stop the column reference moving when the formula is copied across.

## References

Box, G E P and Behnken, D W (1960), Some new three level designs for the study of quantitative variables, *Technometrics*, 2 (4), 455–75.

Carlson, R and Carlson, J (2005), *Design and optimization in organic synthesis*, 2nd ed. (Amsterdam: Elsevier).

Deming, S N and Morgan, S L (1973), Simplex optimization in analytical chemistry. *Analytical Chemistry*, 45, 279A.

Doehlert, D H (1970), Uniform Shell Designs, *Applied Statistics*, 19 (3), 231–239.

Fisher, R A (1935), *The design of experiments*, 1st ed. (London: Oliver & Boyd).

Massart, D L, Vandeginste, B. G. M., Buydens, J. M. C., de Jong, S., Lewi, P. J. Smeyers-Verberke, J. (1997), *Handbook of chemometrics and qualimetrics part A*, 1st ed., vol 20A in *Data handling in science and technology* (Amsterdam: Elsevier).

Mitchell, Toby J (1974), An Algorithm for the Construction of "D-Optimal" Experimental Designs, *Technometrics*, 16 (2), 203–210.

Morgan, E (1991), *Chemometrics: experimental design*, in *Analytical chemistry by open learning* (Chichester: Wiley).

Mullins, E (2003), *Statistics for the quality control chemistry laboratory*, 1st ed. (Cambridge: The Royal Society of Chemistry).

Plackett, R L and Burman J P (1946), The Design of Optimum Multifactorial Experiments, *Biometrika*, 33 (4), 305–325.

Zupan, J and Gasteiger, J (1999), *Neural networks in chemistry and drug design* (New York: Wiley).

# 4

## Quality Control Tools

### 4.1 Introduction

Although thoroughly convinced that no laboratory can function without proper regard to quality in all its myriad forms, the question remains, What do we do? As a quality control manager with a budget and the best aspirations possible, what are the first steps in providing your laboratory or company with an appropriate system (other than buying this book, of course)? Each laboratory is unique, and what is important for one may be less important for another. So before buying software or nailing control charts to your laboratory door, sit down and think about what you hope to achieve. Consider how many different analyses are done, the volume of test items, the size of the operation, what level of training your staff have, whether the laboratory is accredited or seeking accreditation, specific quality targets agreed upon with a client, and any particular problems.

This chapter explains how to use some of the standard quality tools, including ways to describe your present system and methods and ongoing statistical methods to chart progress to quality.

### 4.2 The Concept of Statistical Control

Being in "statistical control" in an analytical laboratory is a state in which the results are without uncorrected bias and vary randomly with a known and acceptable standard deviation. Statistical control is held to be a good and proper state because once we are dealing with a random variable, future behavior can be predicted and therefore risk is controlled. Having results that conform to the normal or Gaussian distribution (see chapter 2) means that about 5 in every 100 results will fall outside ± 2 standard deviations of the population mean, and 3 in 1000 will fall outside ± 3 standard deviations. By monitoring results to discover if this state is violated, something can be done about the situation before the effects become serious (i.e., expensive).

If you are in charge of quality control laboratories in manufacturing companies, it is important to distinguish between the variability of a product and the variability of the analysis. When analyzing tablets on a pharmaceutical production line, variability in the results of an analysis has two contributions: from the product itself and from the analytical procedure. Your bosses are interested in the former, and you, the analyst, must understand and control the latter. It is usually desired to use methods of analysis for which the repeatability is much less than the variability of the product, in which case the measured standard deviation can be ascribed entirely to the product. Otherwise, analysis of variance can be used to split the total variance of duplicate results into its components (chapter 2). In the discussion that follows, the emphasis is on measurement variability, but the principle is the same, and the equations and methods can be used directly to obtain information about the product or manufacturing process.

### 4.2.1  Statistical Process Control

Statistical process control (SPC) is the use of the statistics of the normal distribution to monitor the output of a process (Montgomery 1991). Statistical process control is used largely in the manufacturing industry, where the aim is to reduce defects, or non-conforming items, to as near zero as practical. In the analytical laboratory the output is a result, and, according to the tenets of SPC, this can be monitored in the same way as a ball-bearing, computer component, or automobile. A process is said to be "capable" if the output is within specified tolerances. Take some measure of the output—in an analytical laboratory this could be the difference between the result of the analysis of a certified reference material and the certified value. The material being analyzed should be as near in composition to routine samples as possible, and it should not also be used as a calibration standard, nor should it be used to estimate bias or recovery. The result (call it the error, $\delta$) is expected to have a mean of zero and standard deviation equal to the precision of the analysis ($\sigma$). The tolerance is whatever has been decided to be the maximum permissible error and is designated as ± $T$ about zero (see figure 4.1). (There may be different risks associated with having results that are too high or too low or associated with a nonsymmetrical distribution of results, but for ease a symmetrical tolerance range is assumed.)

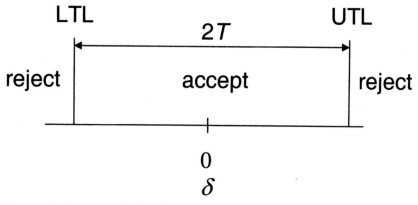

**Figure 4.1.** Tolerance limits for a chemical analysis. $\delta = c_{CRM}(\text{reported}) - c_{CRM}(\text{certified})$. Acceptable limits lie within $\pm T$. LTL = lower tolerance limit. UTL = upper tolerance limit.

iSixSigma provides good web site devoted to SPC, with useful articles and tips on control charts and the like (iSixSigma 2006). NASA (1999) has published a bibliography of SPC, which is a useful start. Software for SPC for MS Excel is also available (McNeese 2006).

A popular approach to comparing the tolerance and standard deviation of measurements is to define the process capability index, $Cp$ or $CI$, as

$$Cp = \frac{2T}{6\sigma} \tag{4.1}$$

Thus, the tolerance range is mapped on to six standard deviations.

If $Cp$ is 1, then 0.3% of results are outside the tolerance levels (i.e., the mean $\pm 3\sigma$ of a normal distribution covers 99.7% of the distribution). In modern manufacturing this is still not acceptable, and $\pm 4\sigma$ is the norm, which leads to a criterion of $Cp > 1.33$ for a capable process. Now only 0.006% of results are not acceptable (see figure 4.2 a). The ultimate goal is to move toward $Cp = 2$ or $\pm 6\sigma$. At the other end, for $1.33 > Cp > 1.00$, some control is necessary; for $1.00 > Cp > 0.67$ the system is considered unreliable; and for $Cp < 0.67$ is unacceptable. A value of $Cp = 0.67$ corresponds to $\pm 2\sigma$ spanning the tolerance range, with an expected failure rate of 5% (see figure 4.2 b). If $\sigma$ is not known, it is estimated from the range of a set of results (see chapter 2, section 2.3.3).

Suppose over a number of results the mean error is not zero. In SPC the mean error as a percentage of the tolerance half-range (which is $T$ when the tolerance is $\pm T$) is known as the index of accuracy, or capability index for setting:

$$C_A = \frac{\delta}{T} \times 100 \tag{4.2}$$

If $C_A$ is < 12.5, the process is considered reliable; between 12.5 and 25 some control is considered necessary; between 25 and 50, the process is unreliable; and if $C_A$ is > 50 the process it is not acceptable.

There is one more index that is used, the overall quality index, or the corrected process capability index, $Cpk$. This is the distance of the mean of the measure to the nearest tolerance limit.

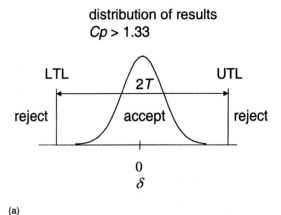

(a)

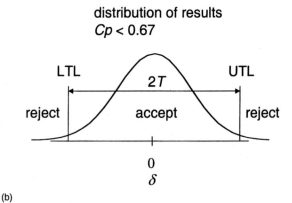

(b)

**Figure 4.2.** Tolerance limits for a chemical analysis. $\delta = c_{CRM}(\text{reported}) - c_{CRM}(\text{certified})$. Overlays are possible distributions of results. (a) Ideal situation with $Cp > 1.33$. Less than 0.006% of results will be outside the tolerance. (b) Poor situation with $Cp < 0.67$. More than 5% of data will be outside the tolerances.

$$Cpk = Cp\left(1 - \frac{\delta}{T}\right) = Cp\left(1 - \frac{C_A}{100}\right) \qquad (4.3)$$

Thus $Cpk$ is $Cp$ corrected for a shift in the mean, measured by $C_A$. The interpretation of $Cpk$ is the same as that for $Cp$.

The approach described here is one way of arriving at a target measurement precision. If a client specifies that results must be within $\pm T$, then the target standard deviation is

$$\sigma_{target} = \frac{2T}{6Cp} \qquad (4.4)$$

which is $T/4$ for $Cp = 1.33$. The difficulty of achieving this is another matter, but this is a useful starting point for a round of continual improvement.

## 4.3 Tools to Describe Non-numerical Information

Before delving into SPC, one needs to have an understanding of the system as a whole. What happens in a laboratory, what variables are important, what variables are likely to need monitoring and optimizing? If you subscribe to any of the popular approaches (total quality management, Six sigma, Crosby, Juran, Deming, Peters, etc.), you will have a shining path to a structured analysis of laboratory operation (Juran and Godfrey 1999). Read and enjoy all of these approaches, and then do what is sensible for your laboratory. It is useful to formally map out what is actually done in the laboratory, down to an appropriate level of detail. There are a number of graphical tools to help (Nadkarni 1991). Some of these are embedded in quality control software, and some come with more general popular spreadsheet and presentation software.

### 4.3.1  Flow Charts

Because a chemical analysis tends to have a linear structure in time, from acquiring the sample through to presentation of the results, it lends itself to graphing as a flow chart. Most operations can be described using the simplest conventions of rectangles containing actions, rounded rectangles with the entry and exit points, and diamonds for decision points that split the chart. Flow charts can be applied to the organization of the laboratory as organizational charts (figure 4.3), work flow charts (figure 4.4), or operations charts for a particular analysis (figure 4.5).

Flow charts can become very complicated. The New South Wales Environmental Protection Agency has a system to aid analysts in the identification of unknown substances (known as Tiphus) that spans 20 full flowcharts that cover every possibility for the analyses required, including treatments

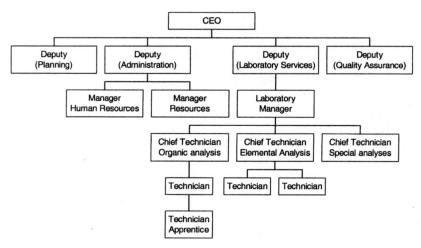

**Figure 4.3.** Organizational chart of an analytical laboratory (Dodgy Laboratories Inc.).

of solids, liquids, and their mixtures. It has recently been put on their local intranet. When computerized, it also acts as an audit of procedures followed by a particular analyst for a particular sample.

### 4.3.2   Cause-and-Effect Diagrams

A cause-and-effect diagram, also known as an Ishikawa diagram or fish bone diagram, is a way of illustrating concepts or activities that impinge on a particular problem. The issue being investigated is placed at the end of an horizontal arrow, then major causes label arrows at 45° joining the main line, and the process continues with sub-causes coming into these lines, and so on (see figure 4.6).

A cause-and-effect diagram can be used as an alternative to a flow diagram, but it is most effective when there is a hierarchy of causes leading to a specified effect. Such a diagram can be used to great effect to identify components of measurement uncertainty (see chapter 6). As an example of troubleshooting a hypothetical problem, consider the liquid chromatography analysis of a raw material. Results of analyses of ostensibly the same material have varied with greater than the expected standard deviation. A first pass might look like figure 4.7, although in a brainstorming session, some of the possible causes may eliminated, perhaps because the answer is known or because that the kind of fault would not cause greater variability but would simply cause a wrong answer.

The use of cause-and-effect diagrams is highly recommended as a tool for structured thinking about a problem. A chemical example is given by Meinrath and Lis (2002).

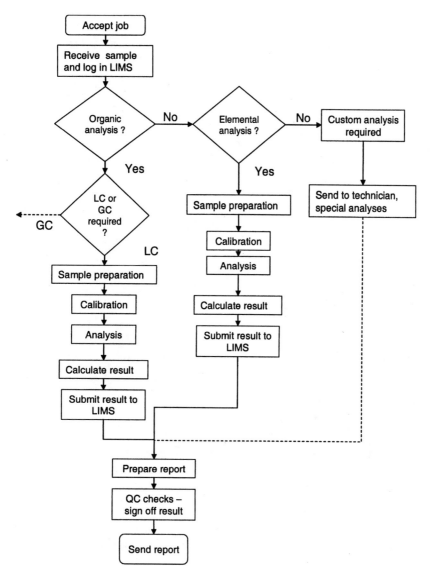

**Figure 4.4.** Flow chart of analytical operations in Dodgy Laboratories Inc.

## 4.4 Graphical Tools for Working with Data

The result of a chemical measurement is a number, with measurement uncertainty and appropriate units. Analysts are accustomed to working with numbers, calibration graphs have been drawn for years, and a laboratory with any kind of commercial success will generate a lot of data. If the results follow a particular distribution, then this statistical knowledge can be used to

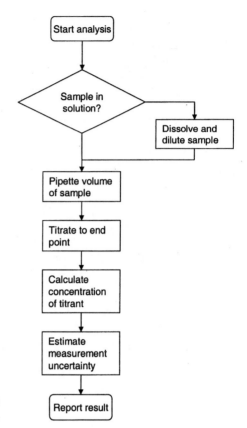

**Figure 4.5.** Flow chart of a simple titration.

predict the behavior of future results and therefore satisfy the laboratory and its clients. A quality control program will monitor the results of analysis of check samples included in each batch. Ideally, a matrix-matched certified reference material will be available that can be taken as a blind sample through the analytical procedure. The mean of a number of analyses of this material allows one to assess bias (or recovery), and the standard deviation of these results can be compared with a target repeatability.

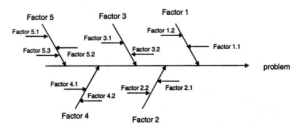

**Figure 4.6.** Generic cause-and-effect (Ishikawa) diagram.

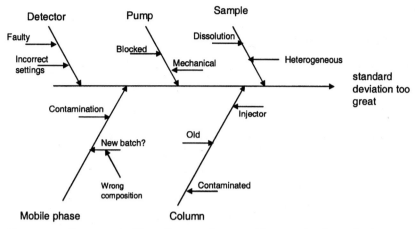

**Figure 4.7.** Cause-and-effect diagram for a problem with a liquid chromatographic analysis.

### 4.4.1  Histograms and Scatter Plots

Histograms are plots of the frequency of the occurrence of ranges of values against the ranges of values. An example is given in figure 4.8, which shows the errors of 75 titration results done by high school students in Sydney in 1997.

Because there were three different solutions being measured, the x-axis is expressed as a percentage error = $100 \times (C_{student} - C_{assigned})/C_{assigned}$ , where

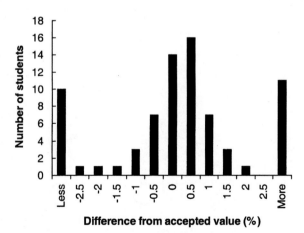

**Figure 4.8.** Histogram of the results (expressed as percent error) of the 1997 Royal Australian Chemical Institute titration competition held in Sydney.

$C_{\text{assigned}}$ is the correct result and $C_{\text{student}}$ is the result reported by the student. There is an immediate impression of a bell-shaped curve around zero with greater numbers at higher and lower errors. In the figure, "less" includes results down to -40% and "more" includes results up to +700%. Twenty-one students (28% of the total) had these extreme results, but the remaining 72% appeared to follow a normal distribution with mean 0.07% and standard deviation 0.84%. It is possible to analyze these data further, but the histogram is a good starting point (Hibbert 2006).

Scatter plots are graphs of one variable against another. A scatter plot is the basis of a calibration graph, but in preliminary data analysis this plot is often used to compare and find correlations in data. As with any method, care must be taken not to be misled by first impressions. The student titrators were in groups of three, each of whom was given a slightly different sample, and there may be an interest in finding out if there is any correlation among their results. Do better schools breed better chemists, for example? If all the results of team member A are plotted against those of team member B, it appears as if this hypothesis may be true (see figure 4.9).

All members of one team arrived at results that were considerably higher than the correct values.[1] This has the unfortunate effect of skewing the whole graph. A correlation coefficient squared is a gratifying, but misleading, 0.994, but if this point is removed along with all the outlying points identified in figure 4.8, the scatter plot of figure 4.10 remains. The $r^2$ is now only .02, and there is clearly no significant correlation.

Which plot is correct? A proper interpretation must consider both graphs and must conclude that the outlying results do show a correlation, possibly due to the fact that some school teams used only one member's value for

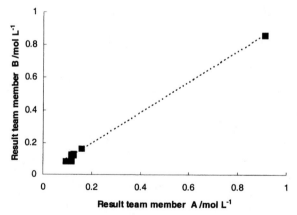

**Figure 4.9.** Scatter plot of the results of team member A against the results for team member B of the 1997 Royal Australian Chemical Institute titration competition held in Sydney.

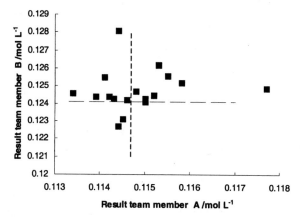

**Figure 4.10.** Scatter plot of the results of team member A against the results for team member B of the 1997 Royal Australian Chemical Institute titration competition held in Sydney, for results that follow a normal distribution (in the "bell" of figure 4.8). Dashed lines show the assigned values of the concentration of each solution.

the concentration of a standard solution. This was measured as a preliminary step (even though they were counseled to each standardize their own solutions), and so any error here would have been manifest in each person's result. Results near the assigned value have only random error, which is not expected to be correlated.

### 4.4.2  Control Charts

For any laboratory that performs a particular activity time and time again, showing the results in a control chart is a good way to monitor the activity and to discover whether a change has caused some deviation in the expected results. In 1924 Walter Shewhart was asked for some kind of inspection report that "might be modified from time to time, in order to give a glance at the greatest amount of accurate information" (Baxter 2002). He duly responded with a sample chart "designed to indicate whether or not the observed variations in the percent of defective apparatus of a given type are significant; that is, to indicate whether or not the product is satisfactory" (Baxter 2002). This was the first control chart, and it has been the basis of statistical quality control ever since. Some averaged measure of the process is plotted over time, and trends or movement outside specifications can be seen. Shewhart was interested in numbers of defects on a production line. The analytical chemists product is a result, and the immediate equivalent to Shewhart's situation would be to repeatedly analyze a certified reference

material and count the number of analyses outside some predefined tolerance. It may be more informative to simply plot the results over time (assuming the same material is analyzed) or, if different reference materials are
used, to plot the frequency of the differences between the result and certified quantity value, as was done with the histogram of the titration results
(figure 4.8). Shewhart overlaid a graph of the results with the limits expected
from a statistical treatment of results and stipulated procedures to follow if
the points went outside these limits. See Woodall (2000) for an overview of
the issues relating to the use of control charts.

### 4.4.2.1  Shewhart Means Chart

The Shewhart means chart is probably the most widely used and can be
easily set up in a spreadsheet. First decide what is being charted—for example, the concentration of a check reference material analyzed twice a day,
in duplicate, as part of the regular batches in the laboratory.[2] The check
samples should not be distinguished from real unknowns, and the order of
their analysis should be randomized. It should be evident why always analyzing the check samples first, or perhaps last, could lead to results that
are not representative of the process. You must select the number of replicates to average ($n$). Because the number of replicates appears in the standard deviation of the mean ($= \sigma/\sqrt{n}$), which is used to set the acceptable
limits of the graph, this decision requires some thought. If $n$ is too small,
the control limits might be too wide to quickly pick up changes in the process, but if it is too great, then individual outliers might become lost in
the average and not picked up. The cost of replicate measurement must
also be considered, and in an analytical laboratory duplicate or triplicate
measurements are often chosen. Grouping may also be done over an ex-
tended time. Perhaps morning and afternoon measurements of the certified reference material are combined to give a daily average, or single
measurements made each day are averaged for a weekly average. The choice
depends on what knowledge is required about the performance of the laboratory and on the timescale.

A Shewhart means chart is a graph of the mean of the replicates against
time (see figure 4.11). Overlaid on the chart are five lines, a center line at
the mean of long-term measurements, two lines equally spaced above and
below the center, called the upper and lower warning limits, and a two more
lines outside the warning lines called upper and lower control (or action)
limits. There are several methods for deciding where the lines go, and the
choice should be made by considering risk and the associated cost of having out of specification results versus stopping the process and inspecting
what is going on. Here is the most simple approach. Assume a certified reference material has been analyzed 4 times each day for 20 days. The mean
for each day ($\bar{x}$), and global mean ($\bar{\bar{x}}$) and standard deviation ($s$) of all the
results are calculated, and the chart may now be drawn.

1. Plot the daily mean ($\bar{x}_i$) for each of the daily results against day.
2. Draw a line at the global mean ($\bar{\bar{x}}$).
3. Draw warning lines at $\bar{\bar{x}} + 2 \times s/\sqrt{4}$ and $\bar{\bar{x}} - 2 \times s/\sqrt{4}$.
4. Draw action lines at $\bar{\bar{x}} + 3 \times s/\sqrt{4}$ and $\bar{\bar{x}} - 3 \times s/\sqrt{4}$.

By plotting the average of $n$ results (here 4), the lines are based on the standard deviation of the mean, which is $s/\sqrt{n}$.

The analyst might have elected to follow the process more closely and plot the duplicate means twice a day. Now the warning and action lines are at $\pm 2 \times s/\sqrt{2}$ and $\pm 3 \times s/\sqrt{2}$, respectively (figure 4.12). Comparing figures 4.11 and 4.12, although the lines are wider apart with fewer repeats, the points are more scattered, too.

The rationale and statistics of the warning and control limits are as follows. For averages of $n$ values of a normally distributed random variable with mean $\mu$ and standard deviation $\sigma$ the means will also be normally distributed with mean $\mu$ and standard deviation $\sigma/\sqrt{n}$ (see chapter 2). Because of the properties of the normal distribution, 95.4% of all values lie between $\pm 2$ standard deviations of the mean, and 99.7% of all values lie between $\pm 3$ standard deviations. If quality control procedures require noting the 4.6% of cases that, although part of the normal distribution, lie outside $\pm 2\sigma$ and require that the analysis be stopped and investigated for the 0.3% of cases outside $\pm 3\sigma$, then this chart is extremely helpful. This is done because, apart from the false alarms (Type I errors) given by these percentages (4.6% and 0.3%), genuine outliers that really do require action are discovered.

Under certain circumstances the system is deemed to be out of statistical control. In this case the analysis must be stopped and the causes of the problem investigated. There are eight indicators of a system being out of

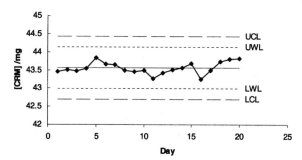

**Figure 4.11.** Shewhart means plot of the duplicate analysis of a certified reference material, twice per day for 20 days. Each point is the mean of the day's four results. Warning limits (UWL and LWL) are at the global mean $\pm 2 \times s/\sqrt{4}$ and control (action) limits (UCL and LCL) at the global mean $\pm 3 \times s/\sqrt{4}$, where $s$ is the standard deviation of all the data.

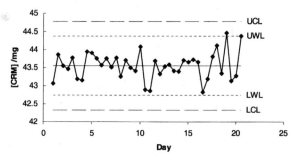

**Figure 4.12.** Shewhart means plot of the duplicate analysis of a certified reference material, twice per day for 20 days. Each point is the mean of one set of duplicate results. Warning limits (UWL and LWL) are at the global mean $\pm$ $2 \times s/\sqrt{2}$ and control (action) limits (UCL and LCL) at the global mean $\pm$ $3 \times s/\sqrt{2}$, where $s$ is the standard deviation of all the data.

statistical control according to the Western Electric rules (Western Electric Corporation 1956), but these indicators can be simplified. The probability of finding this situation given a system that is actually in statistical control (a Type I error) is given in parentheses:

1. One point lies outside a control limit ($P = .003$).
2. Two consecutive points lie between a warning limit and its control limit ($P = .0021$).
3. Seven consecutive points on the same side of the center line ($P = .008$).
4. Seven consecutive points are increasing or decreasing.

How does a chart work? If the system is in control and the true mean coincides with the mean of the chart, then all is well and the frequency of alarms will be given by the probabilities above. But if the mean of the analysis has changed because of an unknown bias, the mean about which the results scatter is $\mu + \delta$. This mean is nearer one set of warning and control limits, so the probability that points will exceed those limits increases. (Of course, the probability that the limits on the opposite side will be exceeded decreases, but overall the chance of observing one of the out-of-control situations is increased.) To illustrate this using the example of figure 4.11, an increase in the true mean by 1 standard deviation of the mean after 10 days has been simulated, and the points now fall around $\mu + \sigma/\sqrt{n}$ (see figure 4.13). The existing warning and control limits remain as they were. The upper warning limit is now 1 standard deviation of the mean away, not 2, the probability of exceeding it is 0.16 (not the 0.023 of an in-control system), and the probability of exceeding the control limit is 0.023 (not 0.003). Action is triggered on day 14 when 2 consecutive means have been above the upper warning limit.

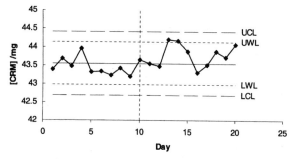

**Figure 4.13.** Shewhart means plot of the duplicate analysis of a certified reference material, twice per day for 20 days. Each point is the mean of the day's four results. A simulated increase in the mean of one standard deviation of the mean is applied after day 10.

Note that had these new points been used to fix the global mean, nothing untoward would have been registered because the center and limit lines would have all moved up together. This shows the importance of choosing the mean and standard deviation correctly. Any data that are used to calculate these parameters must be good. If this sounds like a chicken-and-egg problem, it is, but you should be able to find data you are comfortable with. Alternatively, if measurement is made against a known value, for example a certified reference material, then the center line can be the known value (not the mean of the laboratory's measurement results). In addition, if there is a target measurement uncertainty, then the repeatability component of this measurement can be used as $\sigma$. With time, as the number of quality control data increases, better estimates of the mean and standard deviation can be made. Remember that if you use target measurement uncertainty or other estimates of the standard deviation, unless the system's repeatability happens to coincide with this value, statistical consideration of the $\sigma$ used will not reveal the fractions of results lying outside the limits.

The average run length (ARL) of a chart is, as the term implies, how many points on average must pass before action is triggered given a change in the mean by a given amount. Average run length is the reciprocal of the probability of encountering a change. The Shewhart means chart is very good if the mean suddenly increases by a large amount, but for relatively small changes, which still may be important, it might be a long time before the control limit is violated (table 4.1). Fortunately, the seven consecutive results might show up sooner.

If the standard deviation of the process suddenly increased, then the scatter of the results would increase, and the chance of the system violating the action conditions would be greater. In the example above, the mean has been

Table 4.1. Average run length (ARL) for exceeding an upper or lower control
limit 1/p(total), or giving seven results on one side of the center line 1/p(7) of a
Shewhart means chart for deviations from the mean indicated

| Increase in mean/$(\sigma/\sqrt{n})$ | P (LCL) | P (UCL) | P(total) | ARL= 1/P(total) | P(7) | ARL= 1/P(7) |
|---|---|---|---|---|---|---|
| 0 | 0.001 | 0.001 | 0.003 | 370 | 0.008 | 128 |
| 0.25 | 0.001 | 0.003 | 0.004 | 281 | 0.028 | 36 |
| 0.5 | 0.000 | 0.006 | 0.006 | 155 | 0.076 | 13 |
| 0.75 | 0.000 | 0.012 | 0.012 | 81 | 0.165 | 6 |
| 1 | 0.000 | 0.023 | 0.023 | 44 | 0.298 | 3 |
| 1.25 | 0.000 | 0.040 | 0.040 | 25 | 0.458 | 2 |
| 1.5 | 0.000 | 0.067 | 0.067 | 15 | 0.616 | 2 |
| 1.75 | 0.000 | 0.106 | 0.106 | 9 | 0.751 | 1 |
| 2 | 0.000 | 0.159 | 0.159 | 6 | 0.851 | 1 |
| 2.25 | 0.000 | 0.227 | 0.227 | 4 | 0.918 | 1 |
| 2.5 | 0.000 | 0.309 | 0.309 | 3 | 0.957 | 1 |
| 2.75 | 0.000 | 0.401 | 0.401 | 2 | 0.979 | 1 |
| 3 | 0.000 | 0.500 | 0.500 | 2 | 0.991 | 1 |

LCL = lower control limit, UCL = upper control limit.

returned to its rightful place, but the standard deviation doubled after day
10 (figure 4.14).

The results of the continued simulation are given in figure 4.14. There is
a problem on the first 2 days after the change when both points exceed the
lower warning limit. In both figures 4.13 and 4.14, the changes are quite
obvious on the graph, and an alert quality control manager should be sensitive to such changes, even if action is not taken.

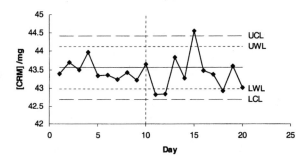

**Figure 4.14.** Shewhart means plot of the duplicate
analysis of a certified reference material, twice per
day for 20 days. Each point is the mean of the
day's four results. A simulated doubling of the
standard deviation is applied after day 10.

### 4.4.2.2  Shewhart Range Charts

As shown in the preceding section, the Shewhart means chart reacts to changes not just in the mean but also to changes in the standard deviation of the results. Nonetheless, I recommend that a separate chart be plotted for the range of the data used to calculate the mean. The range is the difference between the maximum and minimum values, is always positive (or zero), and has a known distribution for normal data. The range is plotted against time, as with the means chart, and the average range across all the data is used to calculate the upper and lower warning and control limits. Perhaps unexpectedly, there are lower limits. If suddenly the range becomes very small the process has changed, although this may be temporarily welcome, it is important to understand what has happened. On the range chart the limit lines are:

Upper control line: $D_{0.001}\bar{R}$
Upper warning line: $D_{0.025}\bar{R}$
Lower warning line: $D_{0.975}\bar{R}$
Lower action line: $D_{0.999}\bar{R}$

where $\bar{R}$ is the global average range, and the parameter $D$ at different probability values is given in table 4.2. Range charts for the two sets of data in figures 4.11 and 4.12 are given in figures 4.15 and 4.16, respectively. As with the means charts, the range charts show all is well with the process.

### 4.4.2.3  CuSum Charts

In terms of breaching the warning and control limits, it will take a long time for a Shewhart means chart to show that the process mean has shifted by a small fraction of the process standard deviation. The ARL for deviation of half a standard deviation from the mean is 155 (table 4.1), so on average 155

Table 4.2.  Parameters for calculating the warning and control limits for ranges as a function of the number of data ($n$)

| $n$ | $D_{0.001}$ | $D_{0.025}$ | $D_{0.975}$ | $D_{0.999}$ |
|---|---|---|---|---|
| 2 | 4.12 | 2.81 | 0.04 | 0.00 |
| 3 | 2.98 | 2.17 | 0.18 | 0.04 |
| 4 | 2.57 | 1.93 | 0.29 | 0.10 |
| 5 | 2.34 | 1.81 | 0.37 | 0.16 |
| 6 | 2.21 | 1.72 | 0.42 | 0.21 |
| 7 | 2.11 | 1.66 | 0.46 | 0.26 |
| 8 | 2.04 | 1.62 | 0.5 | 0.29 |
| 10 | 1.93 | 1.56 | 0.54 | 0.35 |

The average range of the data is multiplied by $D$ to give the lower control limit ($D_{0.001}$), lower warning limit ($D_{0.025}$), upper warning limit ($D_{0.975}$) and upper control limit ($D_{0.999}$). Adapted from Oakland (1992).

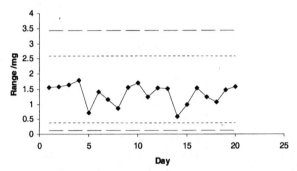

**Figure 4.15.** Shewhart range plot of data in figure 4.11. The range is of the four points taken each day. Dashed lines are control limits and dotted lines are warning limits.

measurements must be made before there is a 50% chance of detecting this error. (Using the full suite of triggers, the condition of 7 results on one side of the mean will be noticed after 13 points.) A more immediate approach to detecting warning and control limits is to plot a CuSum chart, which is a kind of moving average, and is more responsive to persistent but small biases. In a CuSum chart, the cumulative sum of the difference between each mean result ($\bar{x}_i$) and the target value ($x_{target}$) is plotted against time. Thus for each result (the mean of $n$ observations) $x_{target} - \bar{x}_i$ is calculated, keeping the sign. Then in a separate column a running total of the differences is made. Spreadsheet 4.1 presents some hypothetical numbers to show how the accumulation is done.

Note that for a CuSum chart the expected value of the result is used instead of the mean calculated from the data. In an analytical laboratory this

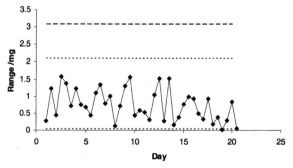

**Figure 4.16.** Shewhart range plot of data in figure 4.12. The range is of the duplicate points taken twice a day. Dashed lines are control limits and dotted lines are warning limits.

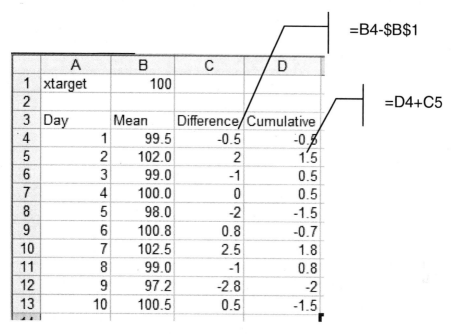

=B4-$B$1

=D4+C5

|    | A | B | C | D |
|----|-----|------|------------|------------|
| 1 | xtarget | 100 | | |
| 2 | | | | |
| 3 | Day | Mean | Difference | Cumulative |
| 4 | 1 | 99.5 | -0.5 | -0.5 |
| 5 | 2 | 102.0 | 2 | 1.5 |
| 6 | 3 | 99.0 | -1 | 0.5 |
| 7 | 4 | 100.0 | 0 | 0.5 |
| 8 | 5 | 98.0 | -2 | -1.5 |
| 9 | 6 | 100.8 | 0.8 | -0.7 |
| 10 | 7 | 102.5 | 2.5 | 1.8 |
| 11 | 8 | 99.0 | -1 | 0.8 |
| 12 | 9 | 97.2 | -2.8 | -2 |
| 13 | 10 | 100.5 | 0.5 | -1.5 |

**Spreadsheet 4.1.** Hypothetical data showing how to calculate the data for a CuSum chart, as it might look in a spreadsheet.

could be the certified value of a reference material. For an unbiased system in statistical control, results are normally distributed about $x_{target}$ and so the expectation of the difference is zero. The data from the spreadsheet are distributed in this manner (see figure 4.17).

A bias will cause an accumulation of differences that can be detected quite early. In the days of graph paper, a V-shaped mask was placed over the graph

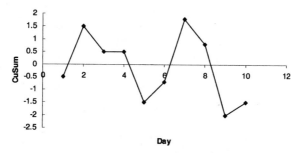

**Figure 4.17.** CuSum chart of the data in spreadsheet 4.1. Data in column D is plotted against the day (column A).

with points outside the V indicating a system out of control. In a spreadsheet, the preferred method of plotting control charts, the following calculations are performed. Two situations are investigated using different calculations, the case of a positive bias and the case of a negative bias. Often the data are clear and only one of these needs to be plotted, but in setting up a spreadsheet you may as well make a column for each. For a positive shift from the mean, the quantity $S_i^+$ is calculated (NIST 2006):

$$S_i^+ = \max\left\{0, S_{i-1}^+ + \frac{\bar{x}_i - x_{target}}{\sigma_{\bar{x}}} - k\right\} \tag{4.5}$$

where $\sigma_{\bar{x}}$ is the standard deviation of the mean, and $k$ is the allowable change in one measurement period expressed in standard deviations of the mean (usually $k = 0.5$). Equation 4.5 shows that only changes greater than $k$ standard deviations contribute to the increase of $S^+$. The standard deviation of the mean can be a target uncertainty or a value calculated from long-run data. Because of its cumulative nature, each value depends on the previous, and so the first in the series ($i = 0$) must be defined as $S_0 = 0$. The equivalent equation for a negative shift is

$$S_i^- = \max\left\{0, S_{i-1}^- - \frac{\bar{x}_i - x_{target}}{\sigma_{\bar{x}}} - k\right\} \tag{4.6}$$

$S_i^+$ and $S_i^-$ are plotted against time, and when either exceeds the threshold $h = 4$, the process is considered out of control.

An example of this scenario is shown in spreadsheet 4.2 and figure 4.18. Here the motor octane number of a standard sample of 95 octane fuel is measured each month in duplicate. The Shewhart means chart never goes over the warning limit, let alone the control limit (see figure 4.19), but something is clearly wrong. The seven-in-a-row rule would be triggered, but CuSum also reveals that the system is out of control at about the same time.

Table 4.3 compares the ARL of a CuSum chart with that of a Shewhart means chart. The CuSum chart picks up even small changes quickly and has a good false positive rate (ARL = 336), and this makes it superior to the run of seven.

A CuSum chart can be used as a general quality control device, even if the system does not go out of control. Inspection of a CuSum chart can be tied to changes in the routine and process because often results in a laboratory are affected by a one-off change. For example, the usual analyst might go on holiday and be replaced by a less (or more) experienced colleague. A new supplier of reagent might be used, or a new instrument might be acquired. Each of these will introduce change in the results that will be sustained until the system reverts to its original condition or changes again.

=C20/SQRT(2)

=AVERAGE(B24:C24)

=D24-$C$19

=E24/$C$21

| | A | B | C | D | E | F | G | H |
|---|---|---|---|---|---|---|---|---|
| 19 | | Target | 95 | | | | | |
| 20 | | Sigma | 2.1 | | | | | |
| 21 | | sem | 1.48 | | | | | |
| 22 | | Cusum k | 0.5 | | | | | |
| 23 | Month | Result 1 | Result 2 | Mean | Mean - target | (Mean - target) sem | S+ | S- |
| 24 | 1 | 96.02 | 96.79 | 96.41 | 1.41 | 0.95 | 0.45 | 0.91 |
| 25 | 2 | 96.14 | 95.70 | 95.92 | 0.92 | 0.62 | 0.57 | 0.00 |
| 26 | 3 | 96.55 | 96.39 | 96.47 | 1.47 | 0.99 | 1.05 | 0.00 |
| 27 | 4 | 97.00 | 96.76 | 96.88 | 1.88 | 1.27 | 1.82 | 0.00 |
| 28 | 5 | 96.60 | 95.57 | 96.08 | 1.08 | 0.73 | 2.05 | 0.00 |
| 29 | 6 | 96.34 | 97.37 | 96.86 | 1.86 | 1.25 | 2.80 | 0.00 |
| 30 | 7 | 96.09 | 97.01 | 96.55 | 1.55 | 1.04 | 3.35 | 0.00 |
| 31 | 8 | 96.64 | 96.24 | 96.44 | 1.44 | 0.97 | 3.82 | 0.00 |
| 32 | 9 | 97.02 | 97.17 | 97.10 | 2.10 | 1.41 | 4.73 | 0.00 |
| 33 | 10 | 97.48 | 97.37 | 97.42 | 2.42 | 1.63 | 5.86 | 0.00 |
| 34 | 11 | 95.55 | 96.28 | 95.92 | 0.92 | 0.62 | 5.98 | 0.00 |
| 35 | 12 | 95.83 | 97.33 | 96.58 | 1.58 | 1.06 | 6.54 | 0.00 |

=MAX(G34+F35-$C$22,0)

=MAX(H34-E35-$C$22,0)

**Spreadsheet 4.2.** Calculation of S+ and S- for the CuSum control chart example of research octane number.

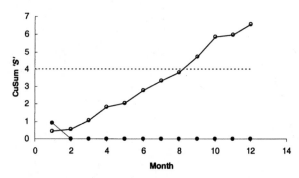

**Figure 4.18.** CuSum S chart of the data of research octane number in spreadsheet 4.2. The system is out of control when *S* exceeds *h* = 4. Open points $S^+$, closed points $S^-$.

Consider some hypothetical data plotted as a means chart in figure 4.20 and as a CuSum chart in figure 4.21. Because the CuSum chart is creating a moving average, the random component is more smoothed out, and the figure detects changes in slope around changes in the mean at observations 10, 19, and 25. The responsiveness of the CuSum chart is due to the fact that when the underlying mean changes, the change adds to every measurement thereafter, which shows up as a clear change in slope of the chart. A Shewhart means chart has to rely on the change being picked up over a number of measurements where each one is affected separately.

One way to use a CuSum chart to understand such changes is to plot a regular analysis of a reference material (which does not have to be a certified reference material, rather a material which is the same and stable over time) and note any obvious changes to the system on the chart. The quality assurance manager can then evaluate any effects due to the changes and take

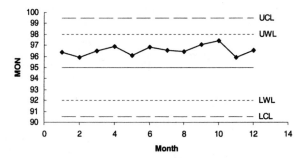

**Figure 4.19.** Shewhart means chart of the means of duplicate data of research octane number in spreadsheet 4.2. The target mean was 95 and the standard deviation was 2.1.

Table 4.3. Average run length for a CuSum chart with $h = 4$ and $k = 0.5$ (both in standard deviations of the mean)

| Standardized mean shift | ARL for $h = 4$ | Shewhart ARL for exceeding control limits |
|---|---|---|
| 0 | 336 | 370 |
| 0.25 | 74 | 281 |
| 0.5 | 27 | 155 |
| 0.75 | 13 | 81 |
| 1.0 | 8 | 44 |
| 1.5 | 5 | 15 |
| 2 | 3 | 6 |
| 2.5 | 3 | 3 |
| 3 | 2 | 2 |

Shewhart ARLs for exceeding the control limits are also shown (see also table 4.1).

appropriate action or start discussions in the quality committee. Realizing that every time a new calibration solution is made the results change somewhat, even if the effect does not trigger wholesale suspension of the process and serious scrutiny, it might lead to concern about measurement uncertainty and a desire to minimize the effect.

### 4.4.2.4 Plotting Charts in Excel

If a laboratory subscribes to statistical software, it is likely that control charts will be available and the data just need to be entered. When using a regular spreadsheet, there are a couple of tips for drawing the limit lines that might

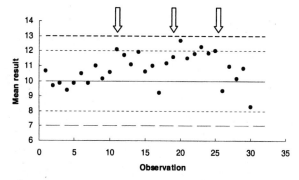

**Figure 4.20.** Shewhart means chart of data that change their mean (arrows) at observations 10 ($y_{target} +1\sigma$), 19 ($y_{target} +2\sigma$), and 25 ($y_{target}$). The target mean is shown at $y_{target} = 10$, and warning and control lines are at $\pm 2\sigma$ and $\pm 3\sigma$.

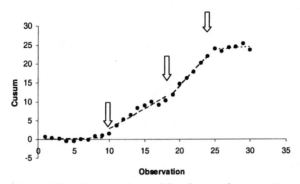

**Figure 4.21.** CuSum chart of the data in figure 4.20.

be helpful. If you want to draw a horizontal line at a value that is computed on the spreadsheet, then there is no need to create columns of identical numbers to be plotted like an ordinary series. For example, to draw a line at the upper control limit, with the chart in focus, but no particular points highlighted, type =SERIES("UCL",(cell with start time, cell with end time), (cell with UCL, cell with UCL),1) and press Enter. An example might be =SERIES("UCL",(B2, B10), (D5, D5),1). A colored maker will appear on the chart at the start and end times. Double click on one of the markers to give the dialogue box for the Format Data Series, and, on the Patterns tab, switch off markers and switch on a line with suitable color and pattern. This series can be created just as well from the menu bar Chart>Source Data and then from the dialogue box Series tab, Add, and within x-values type "= (", then click on the cell with the start time, comma, and then click on this cell again followed by closing parenthesis. Similarly, choose the cell containing the y-values. This method is much better than trying to add a drawing line at about the correct value on the y-axis. As soon as either axis scale changes, the line will be stuck in the wrong place. (And if you forgot to highlight the chart before drawing the line, then it will not even move when the chart is repositioned.)

If you want to put the numbers in directly, rather than refer to cells containing the numbers, use curly brackets. For example, =SERIES(, {0,10}, {0,10},1) puts points at 0,0 and 10,10 and causes a diagonal to be drawn across a chart when a line pattern is switched on. =SERIES(, {0,10} , {1,1},1) puts two points at $x = 0$, $y = 1$ and $x = 10$, $y=1$. Double click on a point, deselect points, and choose a line style, color, and weight to draw the horizontal line.

### 4.4.3 Pareto Diagrams

Pareto diagrams (see figure 4.22) are used to compare magnitudes of quantities and share their name with the Pareto principle, which states something akin to "80% of the work is done by 20% of the people." In this statement, "work" can be replaced by "wealth," "measurement uncertainty," "savings," or "gross domestic product," and "people" can be replaced by "country," "effect," or "contributing component." In quality control, the Pareto diagram is used to describe the relative contributions to measurement uncertainty. a Pareto chart has two parts. A vertical bar chart gives the contributions of the components being displayed, usually expressed as a percentage of the whole, in descending order along the $x$-axis. On top of this, a line with the cumulative total increases from the top of the first bar to 100%. The example, which appears to follow the 80:20 rule, is the components of the uncertainty of the $^{31}P$ nuclear magnetic resonance analysis of the agricultural chemical Profenofos (Al-Deen et al. 2002). Intralaboratory precision has the greatest uncertainty, followed by the purity of the standard (trimethylphosphate) with weighings and molecular weights a long way behind. The actual contribution to the combined standard uncertainty shows an even greater disparity, as the effects are squared and added.

## 4.5  Quality Control Strategies in the Laboratory

Procedures carried out in the laboratory, as opposed to proficiency testing or other interlaboratory collaborations, are known as in-house or internal quality control procedures. When running batches of samples with calibration solutions and unknowns, there are a number of extra samples that can be analyzed that cover different aspects of quality control (QC samples). These QC samples should be documented in the quality manual and be part

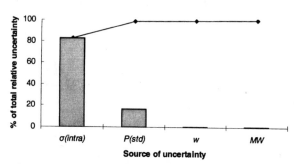

**Figure 4.22.** Pareto chart of the contributions to the uncertainty of the quantitative NMR analysis of Profenofos. The effects are σ (intra), the intralaboratory precision; $P$(std), the purity of the proton standard; $w$, weighings of unknown and standard; MW, the molecular weights of unknown and standard. (Data kindly supplied by T Saed Al-Deen.)

of the test procedure. If there is no particular place in a batch for a given QC sample, these should be randomly analyzed.

### 4.5.1 Blanks

A blank is a sample that does not contain analyte. It might be pure solvent used to establish a baseline for an instrument, or it might be a complex sample matrix. The purpose of a blank is to confirm the absence of analyte. Completely or partially analyzing a blank will reveal contamination of samples during collection, storage, preparation, and analysis. Frequently analyzing blanks (sometimes known as washout samples) will also detect whether there has been carry over of analyte from a previous sample that might have adsorbed on containers, transfer tubing or instruments. When measurements are close to the detection limit, incorporating suitable blanks become more important.

A reagent blank is used to estimate the blank contribution to the instrument response (i.e., the baseline) and is not subjected to all the sample preparation procedures. It is usually the solvent or carrier used to introduce the sample into the instrument. A method blank contains all the components of a sample except the analyte, and it is taken through all steps of the analytical procedure, including digestion, filtration, preconcentration, and derivatization. When samples are prepared and analyzed in batches there should be at least one method blank per batch of samples. For quality control purposes, the method blank is a more complete check of lack of contamination, and the results of this analysis should be used to assess batch acceptance. A batch can be accepted if the method blank is below the detection limit (or limit of quantitation). If analyte is detected at a significant level, the source of contamination must be investigated and measures taken to eliminate or minimize the problem. Correction for a blank measurement should be a last resort, and the uncertainty of the correction must be included in the measurement uncertainty. A regulatory body might set the acceptance criterion for a method blank. If not, the acceptance criterion should be set by the laboratory and documented in its quality manual.

For environmental samples a field blank is similar to a method blank, but it has been exposed to the site of sampling. A field blank is used to screen for contamination during on-site sample collection, processing, sample handling, and transport. For example, in air sampling through a filter that is digested and analyzed in the laboratory, a field blank would be a filter from the same package with the collected filters. The blank filter would be placed in the equipment, and then, without operating the air pump, removed and retained in the same manner as a real sample. In the laboratory the sample would be treated as if it were a routine sample, with the laboratory information management system identifying it as a field blank. In forensic samples where, for example, DNA contamination has been the savior of many a defense case, a blank swab taken to scene and then processed with other samples can be used to argue for a clean process.

### 4.5.2 Replicates

In chapter 2 I introduced the statistics of repeated measurements. Here I describe how these statistics are incorporated into a quality control program. In a commercial operation it is not always feasible to repeat every analysis enough times to apply $t$ tests and other statistics to the results. However, validation of the method will give an expected repeatability precision ($s_r$), and this can be used to calculate the repeatability limit ($r$), the difference between duplicate measurements that will only be exceeded 5 times in every 100 measurements.

$$r = 2.8 \times s_r \qquad (4.7)$$

Any replicated samples, whether QC materials or regular samples, should be monitored by the system, and the sample should be flagged if the difference between duplicates exceeds $r$. Remember that this limit should be exceeded 5 times in 100, so a single result outside the range need not be a cause for immediate concern. On one hand, if no results were flagged, this would be a problem because the repeatability precision in the laboratory does not reflect the reported method precision. On the other hand, this might be good, showing that you are a very careful and precise analyst, in which case a revised repeatability should be calculated from control chart data (see section 4.2). However, a process that has gone wrong and gives the same answer for all samples would also appear to have excellent precision. ISO 5725 (ISO 1994) recommends a third measurement which has its own acceptance criterion, and if that fails further action is warranted.

In some sectors the difference between repeated samples is expressed as the relative percentage difference (RPD), which is defined as

$$RPD = \frac{100 \times (x_{max} - x_{min})}{\bar{x}} \qquad (4.8)$$

In other words, the range is expressed as a percentage of the mean. (With only duplicate results the range is just the difference between the two results.) If a laboratory uses the RPD, then an acceptance criterion is set in the standard operating procedure stating what to do if the criterion is exceeded.

### 4.5.3 Added Control Materials

A variety of reference materials can be added to a sample or analyzed with a batch of samples. These are known in terms of their function.

### 4.5.3.1  Matrix Spike Samples

To assess recovery in samples with complex matrices requiring extensive processing, a traceable reference material is added to a sample, which is then taken through the analysis. Sometimes this is done in duplicate. The recovery is defined as

$$R = 100 \times \frac{\left(x_{spiked} - x_{unspiked}\right)}{x_{added}} \tag{4.9}$$

where $x_{unspiked}$ and $x_{spiked}$ are the measurement results before and after adding a known amount of analyte ($x_{added}$). The spike can be used to assess recovery for each sample, if the matrix varies greatly from sample to sample, or to check a recovery determined as part of the method validation or verification. The amount of spike chosen should give a measurable increase and still lie within the calibration range of the method. It is important not to disturb the matrix, so a spike is usually added as a solid or concentrated solution. Adequate homogenization is then an issue, which you should explore during the validation of the method. The spike material should be an independent sample, not be a calibration solution, and if a correction is to be made for recovery (even if after measurement it is decided to make no correction), its quantity value must be traceable. This is necessary to maintain the unbroken traceability chain.

### 4.5.3.2  Surrogate Samples

A surrogate is a compound not usually found in a sample that is similar in physical and chemical properties to the analyte. When analyzed, a surrogate should be distinguished from the target analytes but otherwise behave as closely as possible to the analyte in terms of recovery and response of the instrument. The behavior of a surrogate is characterized during method validation. One to four surrogates can be added to each blank and sample immediately before sample preparation. Surrogates are particularly useful in chromatography and are chosen to appear across the chromatogram. Typical organic surrogates are compounds that have been isotopically labeled or that have non-reactive groups added such as fluorine or bromine. A surrogate is used as a general quality control material, not for estimating recovery or for calibration. Typically the analysis of a surrogate is plotted in a control chart and trends, and outliers from the average instrumental response are considered in the usual way (see section 4.2).

For environmental analyses in which the matrix varies from sample to sample, the analysis of a surrogate gives a warning if a particular sample is not behaving as expected fashion. As with all QC materials, there must be criteria for the acceptance of surrogate analyses. When there is more than one surrogate, whether all or some fail the criteria is a useful diagnostic. If all fail, this indicates a gross processing error or problematic sample matrix. If only one surrogate has problems, then the analyst needs to decide if this is peculiar to the surrogate or if the concentrations of analytes in the sample are suspect.

### 4.5.3.3  Internal Standards

Internal standards are sometimes used in chromatography (especially with mass spectrometry detection) to quantitatively adjust for variability during a run. The internal standard, like a surrogate, is a compound expected to behave similarly to the analyte but which is not usually present in a sample. If a known and constant amount of the internal standard is added to each measured sample (including calibration solutions, blanks, and QC material), the ratio of the analyte to the internal standard is taken as the instrumental response ($y$). For mass spectrometric detection, isotopically labeled internal standards are preferred, whereas in gas chromatography chlorinated analogues often have similar retention times and detector response. A typical method of employing an internal standard is to add a known and constant amount of internal standard to each sample just before presenting it to the instrument. The ratio of the peaks for the analyte and internal standard is then used for calibration and to measure the unknown. Use of an internal standard can reduce the repeatability of a chromatographic analysis from tens of percent to a few percent.

Internal standards are also used in trace metal analysis by inductively coupled plasma atomic emission spectrometry (ICP-AES) and inductively coupled plasma mass spectrometry (ICP-MS) techniques. An internal standard solution is added to ICP-MS and ICP-AES samples to correct for matrix effects, and the response to the internal standard serves as a correction factor for all other analytes (see also chapter 2).

### 4.5.3.4  Calibration Verification Standards

An initial calibration verification standard should be measured after calibration and before measuring any sample. A calibration verification standard is a standard solution or set of solutions used to check calibration standard levels. The concentration of the analyte should be near either the regulatory level of concern or approximately at the midpoint of the calibration range. These standards must be independent of the calibration solutions and be prepared from a stock solution with a different manufacturer or manufacturer lot identification than the calibration standards. An acceptance criterion is set, usually as a maximum allowable percentage variation (e.g., 5%, 10%). The calibration can be continually verified using either a calibration standard or the initial calibration verification standard. Acceptance criteria must be set and action taken when results fall outside the limits (i.e., stop the analysis, investigate, correct the problem and rerun samples run between the verification standards that were not limits).

### 4.5.3.5 Interference Check Standards

An interference check standard is a standard solution used to verify an accurate analyte response in the presence of possible interferences from other analytes present in the samples. For methods that have known interference problems arising from the matrix or that are inherent in the method, such as ICP-AES (spectral interference lines) and ICP-MS (isotope combinations with similar masses to analyte), these solutions are used in the batch. The interference check standard must be matrix matched to acid content of the samples. Acceptance criteria are set—for example, the magnitude of uncorrected background and spectral interference must not be greater than a stated value.

## Notes

1. A chemical metrologist might wonder how the RACI assigned the correct answer. The titrations were performed a number of times by at least two senior chemists, and the means of the student results that followed normal distributions were not statistically different from the assigned values. Furthermore, the judges decision was final.

2. Already I am assuming that there is enough activity on the analytical production line to warrant this level of checking. For a more modest laboratory a sample may only be measured weekly, or at some other interval.

## References

Al-Deen, T Saed, Hibbert, D B, Hook, J, and Wells, R (2002), Quantitative nuclear magnetic resonance spectrometry II. Purity of phosphorus-based agrochemicals glyphosate (N-(phosphonomethyl)-glycine) and profenofos (O-(4-bromo-2-chlorophenyl) O-ethyl S-propyl phosphorothioate) measured by 1H and 31P QNMR spectrometry. *Analytica Chimica Acta,* 474 (1–2), 125–35.

Baxter, P (2006), 'SPC Analysis unraveled', http:/softwaredioxide.com .community/paper/SPC_All.pdf, (Accessed 1 October, 2006).

Hibbert, B (2006), Teaching modern data analysis with the Royal Australian Chemical Institute's titration competition. *Aust. J. Ed. Chem.* 66, 5–11.

iSixSigma (2006), iSixSigma web site. http://www.isixsigma.com/ (Accessed 22 January 2006).

ISO (1994), Precision of test methods—Part 1: General principles and definitions, 5725-1 (Geneva: International Organization for Standardization).

Juran, J M and Godfrey, A B (eds.) (1999), Jurans quality handbook, 5th ed. (New York: McGraw Hill).

McNeese, W (2006), SPC for MS Excel. http://www.spcforexcel.com/ (Accessed 1 October 2006)

Meinrath, G and Lis, S (2002), Application of cause-and-effect diagrams to the interpretation of UV-Vis spectroscopic data. *Analytical and Bioanalytical Chemistry,* 372, 333–40.

Montgomery, D C (1991), *Introduction to statistical quality control,* 2nd ed. (New York: Wiley).

Nadkarni, R A (1991), The quest for quality in the laboratory. *Analytical Chemistry,* 63, 675A–82A.

NASA (National Aeronautics and Space Administration) (1999), Statistical process control. http://www.hq.nasa.gov/office/hqlibrary/ppm/ppm31 .htm (Accessed 22 January 2006).

NIST (2006), NIST/SEMATECH e-handbook of statistical methods—6.3.2.3. CuSum control charts. http://www.itl.nist.gov/div898/handbook/ (Accessed 22 January 2006).

Oakland, J S (1992), *Statistical process control* (New York: John Wiley and Sons).

Western Electric Corporation (1956), *Statistical quality control handbook* (Indianapolis, IN: AT&T technologies).

Woodall, W H (2000), Controversies and contradictions in statistical process control, in *44th Annual Fall Technical Conference of the Chemical and Process Industries Division and Statistics Division of the American Society for Quality* (Minneapolis, MN: American Society for Quality), 1–10. Available: http://www.asq.org/pub/jqt/past/vol32_issue4/qtec-341.pdf.

# 5

## Interlaboratory Studies

### 5.1 Introduction

No matter how carefully a laboratory scrutinizes its performance with internal quality control procedures, testing against other laboratories increases confidence in a laboratory's results and among all the laboratories involved in comparison testing. Although without independent knowledge of the value of the measurand it is possible that all the laboratories involved are producing erroneous results, it is also comforting to know that your laboratory is not too different from its peers.

An interlaboratory study is a planned series of analyses of a common test material performed by a number of laboratories, with the goal of evaluating the relative performances of the laboratories, the appropriateness and accuracy of the method used, or the composition and identity of the material being tested. The exact details of the study depend on the nature of the test, but all studies have a common pattern: an organizing laboratory creates and distributes a test material that is to be analyzed to the participants in the study, and the results communicated back to the organizing laboratory. The results are statistically analyzed and a report of the findings circulated. Interlaboratory studies are increasingly popular. Ongoing rounds of interlaboratory studies are conducted by most accreditation bodies; the Key Comparison program of the Consultative Committee of the Amount of Substance (CCQM) is one such interlaboratory study (BIPM 2006). There is a great deal of literature on interlaboratory studies (Hibbert 2005; Horwitz

1995; Hund et al. 2000; Lawn et al. 1997; Maier et al. 1993; Thompson and Wood 1993), and an ISO/IEC guide for the conduct of proficiency testing studies is available (ISO/IEC 1997).

## 5.2 Kinds of Interlaboratory Studies

There are three principal groups of studies: studies that test laboratories (proficiency tests), studies that test methods, and studies that test materials (table 5.1). Laboratories that participate in method and material studies are chosen for their ability to analyze the particular material using the given method. It is not desirable to discover any lacunae in the participating laboratories, and outliers cause lots of problems. The aim of the study is to obtain information about the method or material, so confidence in the results is of the greatest importance. When laboratories are being studied, the method may or may not be prescribed, and in this case outliers are scrutinized very carefully. Often proficiency testing (PT) is done on a regular basis, with material sent to a laboratory for analysis as frequently as once a month. A laboratory can then test itself, not only against its peers in the current round, but also against its past performance.

## 5.3 General Methodology

All interlaboratory studies have a common overall procedure (figure 5.1):

1. The organizing body writes the rules for the study, appoints the organizing laboratory, and invites participants.
2. If required, the test material must be prepared or analyzed so that the quantity value is assigned. For example, in interlaboratory comparisons of elements in water organized by the Institute for Reference Materials and Measurements (IRMM) under the International Measurement Evaluation Programme (IMEP), the test materials were prepared by gravimetry or analyzed by a primary method such as isotope dilution mass spectrometry by an independent laboratory with good metrological credentials. Some analysis is usually performed by the organizing laboratory (e.g., to demonstrate homogeneity and stability of the test material).
3. The organizing laboratory sends test portions to the participating laboratories with instructions and timetable for reporting measurement results.
4. On receipt of the measurement results, the organizing laboratory collates the data and statistically analyzes the results. The treatment of outliers and missing data is important in any type of study.
5. The organizing body distributes a report to the participating laboratories and may make the report public as part of a process of

Table 5.1. Kinds of interlaboratory studies

| Name | Purpose | Comments |
|------|---------|----------|
| *Studies to test laboratories* | | |
| Proficiency testing Laboratory performance study Round-robin | To determine the performance of a laboratory with respect to other participants, and (if available) to an independently assigned reference value | Often employed as part of an accreditation scheme (e.g., to ISO/IEC 17025); usually repeated over an extended period of time |
| Cooperative trial | One-off comparison of laboratory performance | May be for contractual purposes |
| Key Comparisons | Assessment of national capability in analysis of specified materials. | Organized by the CCQM and covers important areas related to trade, the environment, and human health; participation is by NMIs. |
| International Measurement Evaluation Program (IMEP) | To provide direct evidence for the degree of equivalence of the quality of chemical measurements (IMEP 2005) | Organized by IRMM for the European Union Laboratories from around the world participate |
| *Studies to test materials* | | |
| Material certification study | To assign a consensus value to a test material | Used as part of a certification scheme, although this does not ensure traceability |
| *Studies to test methods* | | |
| Collaborative trial Method performance study Method precision study | To provide data for a method validation study | Determines the repeatability and reproducibility precision of a method and, if a CRM is used, the method or laboratory bias |
| Interlaboratory bias study | To determine method bias or laboratory bias of a standard method | Similar to a collaborative trial but with a nominated aim |
| Improvement schemes | Validation of new or improved methods by comparison with fully validated method | Less costly exercise than full validation |

deciding if a method is to be adopted as a standard, or if the study is a Key Comparison. Usually the performance of individual laboratories is not disclosed; codes are used to anonymously label the participants.

6. The organizing body reviews the round of tests and prepares for the next round.

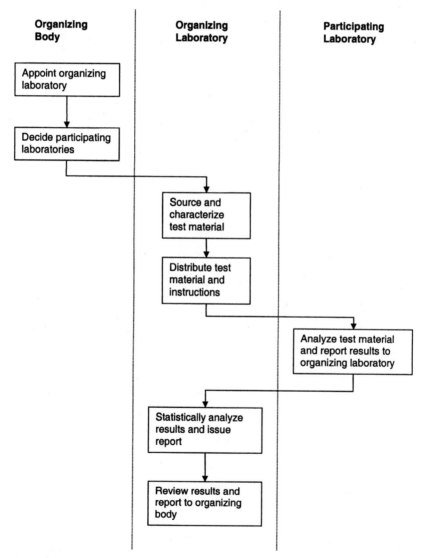

**Figure 5.1.** Schematic of the work flow of an interlaboratory study, showing the sequence of actions by each laboratory or organization.

### 5.3.1  Organizing Structure

There are three kinds of players in most interlaboratory studies: the organizing body, under whose auspices the study is conducted, the referee (organizing) laboratory, and the participating laboratories. The organizing body may be an accreditation body (see chapter 9), an international agency such as the International Bureau of Weights and Measures (BIPM), a national measurement laboratory, a standards organization such as the American Society for Testing and Materials (ASTM) or the International Organization for Standardization (ISO), or the quality assurance section of a large company that wants to compare test results across the laboratories of the company. The organizing body articulates the purpose of the study—for example, to obtain a quantity value for a particular material, to fulfill accreditation requirements of the participants, or to provide evidence of competence for mutual recognition trade agreements. Interlaboratory studies are expensive to set up and maintain, and the cost is borne by the organizing body, often from government funds, or by the participants, as is the case in proficiency studies.

The bureaucrats and politicians who decree an interlaboratory study are not involved in the work of the study. A referee laboratory, often within the organizing body, is tasked with formulating the rules of the study, liaising with the participants, sourcing and characterizing the test material, distributing test materials to participants, receiving and analyzing the results, and issuing the study report. The organizing laboratory must be of the highest competence and totally impartial in its discharge of the duties of the referee. Often the participating laboratories must remain anonymous, and laboratories are identified by a code. Only the participating laboratory will know its code, and participants must be confident that the system will protect their privacy. When the quantity value of the test material must be measured before the study, the organizing laboratory should be able to perform the analysis to an appropriate measurement uncertainty, which will be at least one-tenth that of the reproducibility standard deviation of the participating laboratories. This task is sometimes delegated to national measurement institutes that use primary methods of analysis with small measurement uncertainty and demonstrable metrological traceability.

The participating laboratories might be self-selecting as part of an accreditation program that requires proficiency testing for continued accreditation. For method validation and materials testing studies, laboratories are invited to participate on the basis of their demonstrated competence. Because the results are subjected to statistical analysis, there is a minimum number of laboratories that should take part in a study, below which the statistics are unreliable. It may be prudent to choose more than the minimum number to guard against unexpected outliers, or missing data (e.g., from a laboratory that drops out of the program).

### 5.3.2 Materials

The test material must have the basic properties of identity, stability, and homogeneity.

#### 5.3.2.1 Identity

The test material is chosen to fulfill the aims of the study. In a proficiency testing scheme or a method validation study, the test material is usually as near as possible to typical field samples. There is no advantage in competently analyzing an artificial sample if the same laboratory has difficulty with real samples. The organizing laboratory must know the composition of the test material, and must be sure that the analyte for which a quantity is to be measured is present in about the desired amount. For pure materials this is not a problem, but for natural test materials or complex matrix materials, the organizing laboratory may have to do some analyses before the samples can be sent out to the participants. If the value of the measurand is to be established by an independent laboratory before the study, then the identity requirement is also fulfilled when the measurand is stated.

#### 5.3.2.2 Homogeneity

All bulk material is heterogeneous to some extent. Solutions tend to be more homogeneous than gases and solid mixtures, but the organizing laboratory must demonstrate the homogeneity of the test materials. Sufficient homogeneity is demonstrated when the sampling standard deviation, $\sigma_{sample}$, is much less than the measurement standard deviation of the laboratories in the trial (or target standard deviation in a proficiency testing scheme), $\sigma_p$. The criterion for homogeneity is $\sigma_{sample} \leq \sigma_{allowed}$, where the allowed standard deviation is the maximum permissible and is equal to $0.3(\sigma_p)$. The test involves calculating a critical (95% level) standard deviation based on $\sigma_{allowed}$ and an estimate of the measurement standard deviation $s_{meas}$. The critical value is given by

$$c^2 = \frac{\chi^2_{0.05,m-1}}{m-1}\sigma^2_{allowed} + \frac{F_{0.05,m,m-1}-1}{2}s^2_{meas} \qquad (5.1)$$

where $\chi^2$ is the chi-square distribution, $F$ is the Fisher $F$ distribution, and $m$ is the number of samples analyzed for homogeneity. If $s^2_{sample} \leq c^2$, the sampling variance is acceptable. In Excel the required functions to calculate the critical value $c$ are $\chi^2$ =CHIINV(0.05,m-1), and $F$ =FINV(0.05, m, m-1).

Experimentally, homogeneity is assessed by analyzing in duplicate, using a suitably precise method (with repeatability at least half the target standard deviation of the test), $m$ (at least 10) test samples that have been prepared for distribution. The $2m$ samples should be analyzed in random order to ensure repeatability conditions. Estimates of the sampling variance, $s^2_{sample}$, and the measurement variance, $s^2_{meas}$, are obtained from a one-way ANOVA (see chapter 2), with the $m$ samples being the grouped variable. If the within-group mean square is $\bar{S}_{within}$ and the between-group mean square is $\bar{S}_{between}$, then

$$s^2_{meas} = \bar{S}_{within} \tag{5.2}$$

$$s^2_{sample} = \frac{\bar{S}_{between} - \bar{S}_{within}}{2} \tag{5.3}$$

The denominator 2 in the equation 5.3 reflects duplicate experiments. $s^2_{sample}$ is then compared with the critical value that is calculated from equation 5.1 to determine homogeneity.

### 5.3.2.3  Stability

The test material must be stable because the nature of the analyte and the value of the measurand must be maintained until a laboratory measures it. The organizing laboratory must account for likely transport and storage conditions during the time of testing. In a recent major international program, the Australian result was poor for just one organic compound out of six. Investigation showed that the results appeared to correlate with the distance from the European referee laboratory, and subsequent analysis showed a steady decomposition of that compound. In some long-term studies a great amount of material is made at the outset and used over many years. As long as the material is entirely stable, this approach guards against any variation introduced with the making a new sample each year, but the referee laboratory needs to be sure of the sample's integrity.

### 5.3.3  Statistics

### 5.3.3.1  Identification of Outliers

The organizing laboratory performs statistical tests on the results from participating laboratories, and how outliers are treated depends on the nature of the trial. Grubbs's tests for single and paired outliers are recommended (see chapter 2). In interlaboratory studies outliers are usually identified at the 1% level (rejecting $H_0$ at $\alpha = 0.01$), and values between $0.01 < \alpha < 0.05$ are flagged as "stragglers." As with the use of any statistics, all data from interlaboratory studies should be scrutinized before an outlier is declared.

### 5.3.3.2  Graphical Presentation of Data

Data are often presented as returned results in ranked order, with the mean, robust mean, or assigned reference value for comparison. Some data from a university laboratory practical to determine the fraction of carbonate in a mixture of carbonate and bicarbonate are shown in figure 5.2. The data have been ranked, and the mean ± 2 standard deviations of the set of results are shown. A Grubbs's test on the greatest value (52.2%) gives $g = 1.98$, which is less than the critical value of 2.29 (95%) or 2.48 (99%), so on a statistical basis there is no outlier, but this is due to the large standard deviation (12.4%) of the set. If a $z$ score is calculated based on the assigned value and a target standard deviation of 6% (the experiment requires two titrations, one of which has a difficult end point), analyst 4 would probably be scrutinized (see figure 5.3). Other examples of data analysis and presentation come from the IMEP rounds (IRMM 2006; see section 5.5.1).

Another graphical description of the data is used when comparing the results of several trials is the box plot (also called box-and-whisker plot). A box represents the range of the middle 50% of the data, and whiskers extend to the maximum and minimum values. A line is drawn at the median value. A glance a this plot allows one to assess the symmetry and spread of the data. Figure 5.4 is a box plot for the carbonate data of figure 5.2. Specific plots, such as Youden two-sample plots for method performance studies, are discussed below.

### 5.3.4  Reporting

There are two aspects to a report of an interlaboratory study. First the nature, organization, and treatment of results of the trial must be specified,

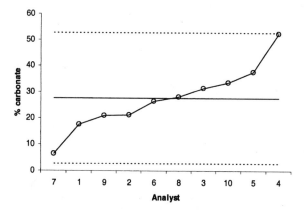

**Figure 5.2.** Plot of ranked data from an analysis of a carbonate–bicarbonate mixture. Lines are drawn at the mean (solid) and ± 2 $s$ (the sample standard deviation of the data) (dotted).

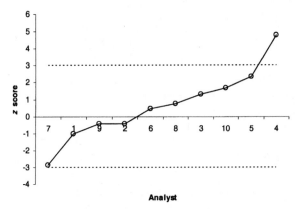

**Figure 5.3.** Data of figure 5.2 plotted as z scores with mean = 23.6% (assigned reference value) and σ = 6.0% (target standard deviation).

and any deviations from the planned procedures should be noted (e.g., missing data). The results should be displayed in a clear manner, with laboratory codes to protect identity if required. The organizing body should write to any laboratories that have been identified in a proficiency test as of concern, and corrective measures (e.g., special monitoring or a site visit to discuss the problems) should begin. The organizing laboratory usually issues

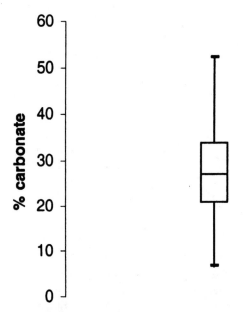

**Figure 5.4.** Box and whisker plot of the data of figure 5.2.

an internal report to the sponsor of the study with a range of details and commentary on the results.

## 5.4  Detailed Protocols

### 5.4.1  Method Performance Studies (Validation)

Standard methods of analysis published by bodies such as the American Society for Testing and Materials (ASTM), Comité Européen de Normalisation (CEN), or ISO are rigorously tested and validated in method performance, or validation, studies. These interlaboratory trials can establish reproducibility and method bias and also give some confidence that the method can be used in different environments. Laboratories are chosen with an expectation that they can competently follow the proposed method, which will have already been extensively validated before an interlaboratory trial is contemplated. To this end, a pilot trial is sometimes undertaken to ensure the method description can be followed and to give an initial estimate of the precision of the method.

#### 5.4.1.1  Trial Procedures

For a full validation trial a minimum of 8 laboratories is recommended, although 15 is considered ideal for establishing reproducibility. Because most methods are used over a range of concentrations, at least five or six samples of concentrations that span the expected range should be analyzed. Duplicate samples should be sent to the laboratories, with either the same concentration or slightly different concentrations ("Youden pairs").

#### 5.4.1.2  Statistical Treatment of Results

One of the aims of a validation study is to establish the repeatability and reproducibility precision. Experiments performed under repeatability conditions (see chapter 2) are those repeated over a short period of time, with the same analyst and same equipment. For an analysis, the repeats must be independent, involving the entire method, and not simply the operation of the instrument several times. Reproducibility conditions are the *sine qua non* of the interlaboratory study. Different laboratories, with different analysts, instruments, reagents, and so on, analyze the sample at different times, but all with the same method. The relationship among these parameters is shown in figure 5.5.

The mean result from the interlaboratory study, $\bar{x}$, follows a model:

$$x_i = \bar{x} + (\delta_i - \bar{\delta}) + \varepsilon_i \qquad (5.4)$$

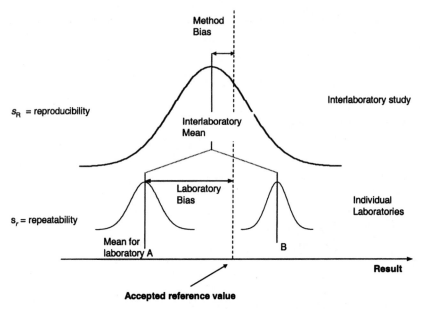

**Figure 5.5.** Diagram showing the relationship between repeatability and reproducibility and laboratory bias and method bias in the context of an interlaboratory study.

where $x_i$ is the value obtained by laboratory $i$, $\delta_i$ is the laboratory bias, $\bar{\delta}$ is the method bias, and $\varepsilon_i$ is the within-laboratory random error (which has an expectation of zero). It is possible to define laboratory bias as the deviation from the method bias, and not from the accepted reference value, although I do not recommend this for validation studies. A more elaborate picture is shown in figure 8.10 where the intralaboratory or intermediate reproducibility is described. This covers the variability over time and in laboratories within which more than one analyst or instrument is regularly used. Across a number of laboratories, $\delta_i$ will average to the method bias with the standard deviation of the results equaling the between-laboratory standard deviation ($s_L$). Together these combine to give the reproducibility standard deviation ($s_R$):

$$s_R = \sqrt{s_r^2 + s_L^2} \tag{5.5}$$

Thus, while an individual laboratory has a bias, $\delta_i$, when combined with the results of other laboratories, this is now a random variable that contributes to the reproducibility.

The above analysis assumes that the results are normally distributed and without outliers. A Cochran test for homogeneity of variance and Grubbs's tests for single and paired outliers is recommended (see chapter 2). Data from laboratories failing these tests at the 99% level ($\alpha = 0.01$) should be rejected, while failure at the 95% level ($\alpha = 0.05$) should cause some scrutiny of the data, but if there are no other counter indications, the data should be retained. If after these tests more than two out nine of the laboratories are excluded, the trial should be halted and no data used. Any outlier should be investigated carefully. I also recommend that only complete data from laboratories be used. If a laboratory reports a missing value, then all the laboratory's results should be excluded.

When duplicate or split samples are sent for analysis, the repeatability and reproducibility can be calculated from an ANOVA of the data with the laboratories as the grouping factor. If the between-groups mean square is significantly greater than the within-groups mean square, as determined by an $F$ test, then the variance due to laboratory bias can be computed as described in chapter 2.

$$s_r^2 = \text{within-laboratory mean square} \qquad (5.6)$$

$$s_L^2 = (\text{between-laboratories mean square} -s_r^2)/2 \qquad (5.7)$$

and then $s_R$ is calculated by equation 5.5.

The reproducibility determined by such a study is often compared with a calculation for the relative standard deviation (RSD) using the Horwitz formula

$$\log_2 (R\%) = 1 - 0.5 \log_{10}(x) \qquad (5.8)$$

where $x$ is the mass fraction of the analyte. A reproducibility that is not near the Horwitz prediction (a so-called Horrat of near unity; see chapter 6) should be scrutinized and a rationale for the discrepancy found. The use of interlaboratory studies and the Horwitz formulas to estimate measurement uncertainty is also discussed in chapter 6.

### 5.4.1.3  Interlaboratory Bias Study

An interlaboratory bias study is a limited form of method performance study used to determine the bias of a standard method or the bias introduced by laboratories that use the standard method. Laboratories are chosen for their competence in performing the method, and the organization is the same as for a method performance study. The number of laboratories in the study is determined by the statistics required. If the bias ($\delta$) is calculated as the difference between accepted reference value and mean of $n$ laboratories' results, the significance can be tested using the standard deviation of the mean, $s_R/\sqrt{n}$. The Student's $t$ statistic is calculated as

$$t = \frac{\delta}{s_R/\sqrt{n}}, \text{ with } n-1 \text{ deg rees of freedom} \qquad (5.9)$$

and the null hypothesis, that the estimated bias comes from a population having mean zero (i.e., there is no bias) is rejected at an appropriately small probability. Knowing the reproducibility from earlier studies, the smallest bias that can be detected by a study can be calculated as a function of $n$, and hence the number of laboratories determined. The standard deviation of the results of the bias study can be tested against the known reproducibility by a chi-square test. If the test reveals a significant difference, the reason for this difference should be investigated, possibly repeating the analyses. Note that testing the bias in this way gives only the probability of a Type I error. At 95% probability a bias will be concluded to be significant 5 times in 100 when it is actually not significant. The probability of an actual bias is not known.

### 5.4.2 Laboratory Performance (Proficiency Testing)

Proficiency testing is increasingly used as part of an accreditation system and used in general to demonstrate the ability of a laboratory to deliver accurate results. Participation in proficiency testing schemes is one of the six principles of valid analytical measurement articulated by the Laboratory of the Government Chemist (see chapter 1). The aim is to determine whether laboratories can achieve a consensus with reasonable dispersion and/or report an accurate result. The latter is only possible if the sample has an accepted reference value against which to compare a laboratory's results. Proficiency testing is used to assess the day-to-day performance of a laboratory, and it should not be used as an opportunity for a laboratory to demonstrate its capacity for higher than average performance. Although it is inevitable that laboratories may take more care with their proficiency testing samples, and perhaps use their proficiency testing results in advertising, the real benefit of proficiency testing is as an aid to quality control. Proficiency testing schemes comprise a continuing series of rounds of testing and are not expected to be a single event. It is the evolving as well as the short term performance of a laboratory that is being assessed.

#### 5.4.2.1 Procedures

Because the laboratories are being tested, and perhaps found wanting, in a proficiency testing scheme, the role of organizer becomes more of judge, jury, and executioner than the collaborative partner of a method validation trial. It is therefore important that absolute probity is maintained by the organizer, which will often be an accreditation body or other official organization. Also, the rules and consequences of participation must be clear. Proficiency testing in itself has nothing to say about accreditation. If an accreditation body decides to take certain actions in response to a laboratory's performance, these actions should be determined and documented in advance. For example, how many times can a laboratory have results outside a z score of ±3 before the accreditation body takes action?

The proficiency testing organization should include an advisory body that has representatives who are practicing chemists, drawn from the sponsor of the program, participants, contractors, and professional organizations. The advisory board should have access to statistical advice. Its job is to oversee testing and make recommendations to the organizing body on matters such as the frequency of testing rounds, the numbers and kinds of samples, documentation and instructions provided to the participants, and treatment and reporting of results.

To ensure fairness for all participants, the rules of proficiency testing schemes are often more exact than other interlaboratory trials. Results must be returned by a specified time and must follow the defined format to be valid. Collusion between laboratories is unprofessional and is against the spirit of testing. To discourage collusion, the organizing body may announce that slightly different materials will be sent out at random to the participants. Where split pairs are sent for analysis, it is possible to design the difference in levels of the pair to maximize the probability of detecting collusion between laboratories (Wang et al. 2005). Because of the use of a standard deviation for proficiency testing, the participants are usually instructed not to return a measurement uncertainty.

### 5.4.2.2 Samples and Methods

Samples distributed for analysis should have known quantity values with estimated measurement uncertainties, although, of course, these are not released to the participants until after the tests. The value of the measurand ($X_{assigned}$) can be assigned by

1. Analysis by an expert laboratory of high metrological quality. This will often be one or more reference laboratories such as a national measurement institute.
2. The value and measurement uncertainty of a certified reference material.
3. A formulation using materials of known composition and certified quantity values.
4. The post-hoc consensus value from the testing round.

Post-hoc consensus does not independently establish the value of the measurand, and although it provides internal consistency, comparison with other testing rounds and proficiency tests should be made with care. In some cases the outliers from the consensus mean in a round may be the only ones with the correct answer. When an empirical method is used to establish the assigned value, this method should be made known to the participants in advance. Where the assigned value is provided with an uncertainty (as it would be in cases 1–3 above), the uncertainty should be small compared to the standard deviation used to calculate the $z$ score (see next section). For unequivocal use of the results, the following condition should hold

$$u_x^2 < 0.1\sigma_p^2 \qquad (5.10)$$

where $u_x$ is the standard uncertainty of the assigned value, and $\sigma_p$ is defined in the next section.

The method of analysis is either prescribed, when this is regulated by law, for example, or the choice is left to the participant. In the latter case the method is reported with the result. In keeping with the aims of a proficiency testing scheme, the method used should be the routine method employed by the laboratory, and not some enhanced protocol designed to improve performance in the scheme.

### 5.4.2.3 Statistical Treatment of Results

In contrast to the use of collaborative trials to estimate repeatability, reproducibility, and bias of a method, where the participating laboratories are treated as random representatives of a population of competent laboratories, in proficiency testing each laboratory is an independent entity that is being assessed for its ability to return an accurate measurement result. In the terms used in ANOVA, the factor "laboratory" in proficiency testing is a fixed effect, while in method validation it is a random effect. Therefore in this case, the within and between variables variance has no useful meaning, particularly of laboratories use different methods, other than to determine if there is significant differences among the group of laboratories.

The results of the laboratories are assessed by converting them to a $z$ score (see chapter 2).

$$z_i = \frac{x_i - X_{assigned}}{\sigma_p} \qquad (5.11)$$

where $x_i$ is the result of laboratory $i$, $X_{assigned}$ is the assigned reference value, and $\sigma_p$ is the assigned standard deviation, known as the standard deviation for proficiency assessment. The variable $\sigma_p$ has also been called the target standard deviation, but this terminology is not now recommended. Because $\sigma_p$ is not necessarily the standard deviation of the normally distributed results of the testing round, it is not possible to ascribe a significance to a particular value. Thus $z = \pm2$ should not have connotations of a 95% probability range. The organizing body should give some guidance as to the interpretation of the $z$ score. Often $\sigma_p$ is chosen to make scores outside $\pm3$ unwanted and subject to scrutiny, but the usual statistical interpretations must not be made without some understanding of the implied distribution. When deciding on a fit-for-purpose standard deviation, a value should be chosen that trades off a few incorrect decisions that might be made as a result of a greater measurement uncertainty and the cost of achieving a lesser measurement uncertainty.

Although independent assignment of the value of the measurand is preferred, many proficiency tests use the consensus mean and standard deviation to calculate the z scores. This is the cheapest method, and it has a kind of egalitarian feel, in which everyone contributes to the assignment of the right answer. To avoid the problem of outliers skewing the normal statistical parameters, a robust mean and standard deviation are calculated. For $X_{assigned}$ the median is chosen (see chapter 2; i.e., the middle value of the data arranged in order of magnitude). The robust standard deviation should be calculated from the MAD (median absolute deviation) or IQR (interquartile range; see also chapter 2).

The data should always be scrutinized carefully. Extreme results might be due to individual laboratory error, but sometimes there are two or more groups of results that correspond to a particular source of bias. Identifying such groups can be very informative and lead to highlighting particular forms of bad practice. However, the presence of groups makes a sensible assignment of quantity value from the results very difficult, if it is impossible to decide which group has the more correct result. In some cases the results will have to be reported in their raw state with no attempt to calculate z scores. When more than one sample has been analyzed, the individual results can be quoted, or a lumped statistic such as the sum of squared z scores can be used (Uhlig and Lischer 1998).

### 5.4.2.4  Reporting Proficiency Tests

The report should contain a clear statement of procedures for dealing with the data and the determination of test statistics (e.g., treatment of missing data and method of calculation of z scores). The identity of the laboratories should be coded. Ideally, all data should be reported to allow a laboratory to check the calculation of its z score. In these days of Internet and spreadsheets, there should be no impediment to providing this information. Graphs of the z scores against laboratory code or ordered z scores or laboratory means are often reported. Where split level samples have been distributed, a Youden plot is given. I do not recommend ranking z-scores in a league table of laboratories. The outcome of proficiency testing is not to highlight winners and losers but to encourage acceptable practice by every laboratory.

### 5.4.3  Materials Certification

Because a consensus value does not guarantee a correct result, the use of interlaboratory studies to establish the quantity value of a would-be reference material must be undertaken with great care. A traceable quantity value per se is not established by the consensus of a number of laboratories, but if each of those laboratories can demonstrate that they have reported a traceable measurement result, the weight of the group of these measurements is compelling evidence that the combined measurement result is an appropriate estimate of the true value. It follows that only laboratories of the highest metrological standards should be used to certify the values of quantities of reference materials.

The organizer of an interlaboratory study to assign a quantity value to a reference material is usually a national or international certifying authority, such as the International Atomic Energy Agency (United Nations), the Community Bureau of Reference (European Union), or a national measurement institute.

### 5.4.3.1  Choice of Materials, Methods, and Laboratories

Because the material will be distributed as a reference material after the study, there must be enough material to satisfy the needs of the test and its future uses. Homogeneity and stability must be demonstrated, and after the first certification round, later rounds can be planned to establish shelf life of the material. Usually one laboratory has the responsibility for performing homogeneity and stability tests. Subsamples are stored at temperatures ranging from −20°C to +50°C and analyzed once a month for a period of 3 or 4 months.

Analysis of a material by a small number of laboratories using the same method runs the risk of introducing a method bias into the result. At least 20 laboratories are chosen for their high standard of analytical competence and their ability to apply different methods, where this is appropriate. If necessary the laboratories will also be asked to use different pretreatment methods.

### 5.4.3.2  Procedure for Materials Certification

A round of preliminary tests establish homogeneity and a likely consensus value. If there is any discrepancy among the results, a second round may be needed. Only those laboratories that achieve reasonable results in the initial rounds participate in the final certification round, although any laboratories that do not achieve consensus should be investigated to make sure they are not the only ones doing the analysis correctly. For the certification round, at least four subsamples, distributed on different days, should be analyzed in duplicate to allow estimation of repeatability ($s_r$) and the effect of the influence factor, time.

### 5.4.3.3  Statistical Treatment of Results

The normality of the distribution of results is checked by an appropriate test, such as the Kolmogorov-Smirnov test, and outlier tests are performed on variances and means. It is important that outliers or other untoward results are properly investigated. The organizers and testing laboratories must be confident about the consensus mean result. ANOVA is used to calculate the different contributions to the variance, between-laboratory reproducibility, and within-laboratory repeatability. If different methods have been used, the method is also investigated as a factor.

The certified value is usually taken as the grand mean of the valid results. The organizer uses standard deviation as the basis for calculating the measurement uncertainty. Results from the laboratories will include their own estimates of measurement uncertainty and statements of the metrological traceability of the results. There is still discussion about the best way to incorporate different measurement uncertainties because there is not an obvious statistical model for the results. One approach is to combine the estimates of measurement uncertainty as a direct geometric average and then use this to calculate an uncertainty of the grand mean. Type A estimates will be divided by $\sqrt{n}$ ($n$ is the number of laboratories), but other contributions to the uncertainty are unlikely to be so treated.

## 5.5  International Interlaboratory Trials

International interlaboratory trials are conducted to assess or demonstrate the ability of laboratories across nations to achieve comparable measurement results. International trials have become of interest as international trade that depends on chemical analysis grows. The IMEP program compares field laboratories, while the Key Comparisons program of the CCQM targets national measurement institutes and is the basis of mutual recognition arrangements of the International Committee for Weights and Measures (CIPM). The goal of these initiatives is make measurements acceptable everywhere, no matter where they are made

### 5.5.1  International Measurement Evaluation Program

In a farsighted move in 1989, the European Union laboratory IRMM started a series of interlaboratory comparisons to provide objective evidence for the degree of equivalence and the quality of chemical measurements by comparing a participant's measurement results with external certified reference values (IRMM 2006). At the time most proficiency testing schemes used consensus results for the mean and standard deviation to derive $z$ scores. With the IMEP-1 analysis of lithium in serum, the world was alerted to the problem of lack of accuracy in analytical measurements. The data of the first IMEP-1 trial are replotted in figure 5.6; notice that the apparent outlier was the only laboratory to come close to the assigned value.

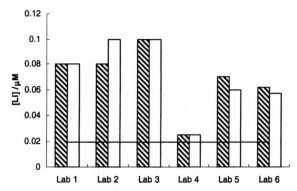

**Figure 5.6.** Results of IMEP-1. Two samples of serum containing 0.019 ± 0.001 mmol L⁻¹ Li were analyzed by six laboratories with the results as shown. (Reproduced with permission from Hibbert and Gooding 2005.)

The results of IMEP-1 clearly pointed to the problem of lack of equivalence of measurement results and demonstrated that studies needed to use independently established quantity values. IMEP operated from the beginning under the auspices and with the support of the International Union of Pure and Applied Chemistry, EURACHEM, the Association of European Metrology Institutes, and Cooperation in International Traceability in Analytical Chemistry.

The methodology of IMEP is the same as for any proficiency testing scheme. The participating laboratories conduct analyses using their routine procedures. The IMEP-certified test sample is well characterized and has reference values certified by laboratories that have demonstrated capability to make the particular measurement. In some cases, such as trace elements in water, the samples are natural samples that have been analyzed by a primary method (e.g., isotope dilution MS), or synthetic samples, in this case high-purity water to which elements have been added in known amounts. Unlike other proficiency testing schemes, participants in IMEP are invited to state uncertainty estimates for their reported results, together with the method used and their self-assessed competence and accreditation status. IRMM has become the leader for these kinds of studies and provides characterized samples for the CCQM Key Comparisons and pilot studies for national metrology institutes worldwide.

The IMEP rounds involve laboratories are from all around the world of different metrological function and experience. Measuring the same quantity in the same sample yields evidence of measurement capability. Complete anonymity of laboratories is preserved, and although the program provides analysis of the results in the form of tables and graphs, laboratories draw their own conclusions about the accuracy of their results.

The analytes and matrices are chosen for their prominence in cross-border trade and environmental or political issues, and they include substances such as rice, wine, water, fuel, and human serum. Table 5.2 is a list of IMEP rounds up to the end of 2006. The majority of analytes are elements. This reflects the history of IRMM, which has been prominent in the measurement of atomic weights and the development of IDMS as a primary method for the analysis of elemental amounts.

Table 5.2. The International Measurement Evaluation Programme (IMEP) rounds up to 2006 arranged by analyte and matrix (IRMM 2006)

| IMEP comparison | Material and matrix | Elements | Years of study |
|---|---|---|---|
| 4 | Bovine serum | Li, Cu, Zn | 1991–1995 |
| 11 | Car exhaust catalysts | Pt, Zr, Ce, Hf | 1998–1999 |
| 8 | $CO_2$ | $n(^{13}C)/n(^{12}C)$ and $n(^{18}O)/n(^{16}O)$ | 1997–1999 |
| 1 | Human serum | Li | 1989 |
| 5 | Human serum | Fe | 1991–1994 |
| 7 | Human serum | Ca, Cl, Cu, Fe, K, Mg, Na, Se, Zn | 1997–1998 |
| 17 | Human serum | Ca, K, Li, Mg, Na, Zn and minor organic constituents | 2000–2003 |
| 2 | Polyethylene | Cd | 1990–1991 |
| 10 | Polyethylene | Cd, Cr, Hg, Pb, As, Cl, Br, S | 1997–1998 |
| 13 | Polyethylene | Cd, Cr, Hg, Pb, As, Cl, Br, S | 1999–2000 |
| 19 | Rice | Cu, Cd, Zn, Pb | 2002–2003 |
| 14 | Sediments | Trace elements | 1999–2000 |
| 21 | Sewage sludge | Metals, PCBs and PAHs | 2005 |
| 18 | Sulfur in diesel fuel (gasoil) | S | 2004–2005 |
| 22 | Sulfur in petrol | S | 2006 |
| 3 | Trace elements in water | B, Ca, Cd, Cu, Fe, K, Li, Pb, Rb, Zn | 1991–1993 |
| 6 | Trace elements in water | Ag, B, Ba, Cd, Cu, Fe, Li, Mo, Ni, Pb, Rb, Sr, Tl, Zn | 1994–1995 |
| 9 | Trace elements in water | B, Ca, Cd, Cr, Cu, Fe, K, Li, Mg, Ni, Pb, Rb, Sr, U, Zn | 1998–1999 |
| 12 | Trace elements in water | Ag, B, Ba, Ca, Cd, Cu, Fe, K, Li, Mg, Mo, Ni, Pb, Rb, Sr, Tl, Zn | 2000–2001 |
| 15 | Trace elements in water | As, B, Cd, Cr, Cu, Fe, Mg, Mn, Ni, Pb | 2001–2002 |
| 20 | Tuna fish | As, Hg, Pb, Se, methyl mercury | 2003–2004 |
| 16 | Wine | Pb | 1999–2001 |

IMEP-12, the study of trace elements in water, followed studies (IMEP-3, 6, 9) that focused on the ability of laboratories to correctly analyze a suite of elements at trace, but not ultra-trace, levels. Because of the general importance of measuring elements in the environment, a large number of laboratories took part (348 laboratories from 46 countries on 5 continents), which provided a spread of experience, methods, and results. Participants of IMEP-12 measured the amount content of the elements As, B, Cd, Cr, Cu, Fe, Mg, Mn, Ni, and Pb in two samples of water. The samples for analysis were subsamples of a single solution that was prepared by gravimetric addition of concentrated mono-elemental solutions in purified water in order to keep the approximate concentration of the elements for measurement close to the legal limits for water intended for human consumption (European Union 1998). Five institutes of high metrological standing analyzed the solutions and provided measurement results for each element and sample and a full measurement uncertainty (GUM; see chapter 6). These institutes were University of Natural Resources and Applied Life Sciences (BOKU Vienna, Austria), IRMM (Geel, Belgium), Federal Institute for Materials Research and Testing (BAM Berlin, Germany), National Measurement Institute Japan (NMIJ Tsukuba, Japan) and the Laboratory of the Government Chemist (LGC Teddington, UK). According to the IMEP policy, if the RSD uncertainty of the value of a measurand is less than 2%, it is deemed certified; if the value is greater than 2%, the result is "assigned."

The participants were free to use whatever methods they preferred. The majority used inductively coupled plasma (ICP; with optical emission or mass spectrometry) or atomic absorption spectrometry (AAS; flame or electrothermal), although 38 different named techniques were used. The results are fully documented in the report from IRMM (Papadakis et al. 2002) and plotted as ordered results with reported measurement uncertainties. There is a great range of results and reported measurement uncertainties. Out of 242 laboratories that returned results for As, I counted (from the graph in the report, so there is an uncertainty here of plus or minus a couple of laboratories) 104 with results in the certified range, and if laboratories whose error bars make it into the certified range are also counted as successful, this becomes 149, or 61% of the total. Thirty-six laboratories (15%) are outside ± 50%. The IMEP studies have also shown great variation in the reported measurement uncertainty. One laboratory that obtained a result that was within the assigned range also reported a measurement uncertainty of greater than ± 50%, thus making the result useless.

An interesting conclusion from many IMEP studies is the lack of correlation between results and (1) method used, (2) accreditation status of the laboratory, (3) reported familiarity with the method, and (4) country or region of origin. In all cases some laboratories obtain reasonable results and some do not. For many studies participation in a series of rounds appears to lead to some improvement, but the one-quarter to one-third of laboratories not obtaining a result that can be said to be equivalent to the certified or assigned value seems to hold in many studies (see chapter 1). One of the leading European chemical metrologists, Professor Paul De Bièvre, has interpreted these observations as an argument for well-trained and committed staff. It could be that funds spent on the latest technology that is claimed to be idiot proof might not be so well placed. The analysis of a sample is much more than knowing which buttons to press on the instrument.

### 5.5.2    Key Comparison Program of CCQM

In 1999, the BIPM started the Key Comparisons program in which countries (member states of the Metre Convention and Associates of the General Conference on Weights and Measures) sign mutual recognition arrangements (MRAs) to accept standards, and, if desired, calibration and measurement certificates issued by other signatories' national measurement institutes. A result of the program is that participating national metrology laboratories can establish the degrees of equivalence of their national measurement standards.

According to the BIPM (BIPM 2006) the objectives of the CIPM MRA are to:

- Provide international recognition of, and to improve the realization of national standards,
- Provide confidence in, and knowledge of the measurement capabilities of participating laboratories for all users, including the regulatory and accreditation communities,
- Provide the technical basis for acceptance between countries of measurements used to support the trade of goods and services— "equivalent" certificates issued in the framework of the MRA, which can be accepted worldwide,
- Reduce technical barriers to trade arising from lack of traceability and equivalence.

To maintain membership of the mutual recognition arrangement, a national measurement institute must take part in rounds of the Key Comparisons. The Key Comparisons organized by the CCQM have consisted of a wide variety of matrices and measurands. As of 2006, 80 key comparisons covering all areas of chemical measurement had been completed. As with IMEP, the attempt is to cover all important areas that are involved in international commerce, but the Key Comparisons program covers a wider range of analytes and matrices. Examples of areas include health (cholesterol in fish), food (arsenic in fish), environment (gases in air), advanced materials (semiconductors), commodities (sulfur in fossil fuels), forensics (ethanol in air/ breathalyzer), biotechnology (DNA profiling), and general analysis (pH).

#### 5.5.2.1    An Example: CCQM-K6

In 2001, a comparison was undertaken to assess the capability of countries to measure cholesterol in human serum (Welch et al. 2001). After a pilot

study, seven national measurement institutes took part in a round to analyze two samples of serum provided by the National Institute for Standards
and Technology. All the laboratories in the study used an isotope dilution/
GCMS method, which involves adding a known mass of a cholesterol material with a stable isotope label to a known mass of serum. Esters of cholesterol are hydrolyzed to cholesterol, which is converted to a trimethylsilyl
derivative to improve separation and detection. The ratio between the native cholesterol and the isotopically labeled material that was added is
measured by GCMS. The cholesterol content of the serum is measured by
comparing this ratio with that of known calibration mixtures of the same
labeled cholesterol and unlabeled cholesterol of known purity. The results
for one material are graphed in figure 5.7, together with the assigned value
and expanded uncertainty.

Of importance to the traceability of the results is a proper estimate of the
measurement uncertainty of each participant. The error bars in figure 5.7
are the expanded uncertainties reported by the laboratories, in some cases
with a coverage factor relating to more appropriate degrees of freedom. Table
5.3 gives the uncertainty budget from one of the participants, the National
Analytical Reference Laboratory, now part of the Australian National Measurement Institute. The repeatability of the measurements of standard and
test material contributes the greatest uncertainty.

### 5.5.2.2  Degree of Equivalence

A measure of the agreement between results from two laboratories, or between
a laboratory result and the assigned value, has been developed in the Key

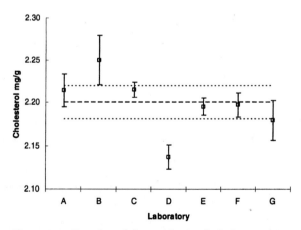

**Figure 5.7.** Results of the analysis of cholesterol in
serum in a round of the CCQM Key Comparisons
(CCQM-K6). The laboratories are identified in the
KCDB report (Welch et al. 2001).

Table 5.3. Measurement uncertainty budget for the measurement of cholesterol in serum by isotope dilution/GCMS

| Source of uncertainty | Type | Relative uncertainty (%) | Degrees of freedom |
|---|---|---|---|
| Initial mass sample solution | B | 0.008 | $\infty$ |
| Final mass sample solution after dilution | B | 0.0004 | $\infty$ |
| Mass sample solution for blend | B | 0.008 | $\infty$ |
| Mass spike solution for blend | B | 0.014 | $\infty$ |
| Mass standard for calibration blend | B | 0.008 | $\infty$ |
| Mass spike for calibration blend | B | 0.014 | $\infty$ |
| Concentration of calibration solution | B | 0.06 | $\infty$ |
| Repeatability of cholesterol/$^{13}C_3$-cholesterol ratio for sample blends | A | 0.426 | 5 |
| Repeatability of cholesterol/$^{13}C_3$-cholesterol ratio for standard blends | A | 0.397 | 6 |
| *Calculation of expanded uncertainty* | | | |
| Combined RSD % | | 0.59 | 10 |
| Coverage factor $k$ | | 2.28 | |
| Relative expanded uncertainty (%) | | 1.29 | |
| Mean of results | | 2.250 mg/g | |
| Expanded uncertainty ($U$) | | 0.029 mg/g | |

Data recalculated from Welch et al. (2001).

Comparisons program and is known as "degree of equivalence." This consists of the signed difference between the results, with a combined expanded uncertainty. Thus, for two laboratories with results $x_1 \pm U_1$ and $x_2 \pm U_2$, the degree of equivalence is $(x_1 - x_2)$, with expanded uncertainty $\sqrt{U_1^2 + U_2^2}$. A Student's $t$ test with $H_0$ that the results come from a population with degree of equivalence = 0 can then be used to determine if there is a significant difference between the results.

## References

BIPM (2006), The BIPM Key Comparison Database (KCDB). Available: www.bipm.org/kcdb.

European Union (1998), Council Directive on quality of water intended for human consumption, 98/83/EC. Brussels: European Union.

Hibbert, D B (2005), Interlaboratory studies, in P J Worsfold, A Townshend, and C F Poole (eds.), *Encyclopedia of analytical science*, 2nd ed., vol. 7 (Oxford: Elsevier), 449–57.

Hibbert, D B and Gooding, J J (2005), *Data analysis for chemistry: an introductory guide for students and laboratory scientists* (New York: Oxford University Press).

Horwitz, W (1995), Revised protocol for the design, conduct and interpretation of method performance studies. *Pure and Applied Chemistry*, 67, 331–43.

Hund, E, Massart, D L, and Smeyers-Verbeke, J (2000), Inter-laboratory studies in analytical chemistry. *Analytica Chimica Acta*, 423, 145–65.

IRMM (2006), The International Measurement Evaluation Programme. Available: http://www.irmm.jrc.be/imep/ (Geel: Institute for Reference Materials and Measurements).

ISO/IEC (1997), Proficiency testing by interlaboratory comparisons—Part 1: Development and operation of proficiency testing schemes, Guide 43 (Geneva: International Organization for Standardization).

Lawn, R E, Thompson, M, and Walker, R F (1997), *Proficiency testing in analytical chemistry* (Cambridge: Royal Society of Chemistry).

Maier, E A., Quevauviller, Ph, and Griepink, B (1993), Interlaboratory studies as a tool for many purposes: proficiency testing, learning exercises, quality control and certification of matrix materials. *Analytica Chimica Acta*, 283 (1), 590–99.

Papadakis, I, Aregbe, Y, Harper, C, Nørgaard, J, Smeyers, P, De Smet, M, et al. (2002), IMEP-12: Trace elements in water, report to participants (Geel: European Commission – Joint Research Centre, Institute for Reference Materials and Measurements).

Thompson, M and Wood, R (1993), The international harmonized protocol for the proficiency testing of (chemical) analytical laboratories, *Pure and Applied Chemistry*, 65, 2123–44.

Uhlig, S and Lischer, P (1998), Statistically-based performance characteristics in laboratory performance studies. *Analyst*, 123, 167–72.

Wang, W, Zheng, J, Tholen, D W, Cao, Z, and Lu, X (2005), A statistical strategy for discouraging collusion in split-level proficiency testing schemes. *Accreditation and Quality Assurance*, 10 (4), 140–43.

Welch, M J, Parris, R M, Sniegoski, L T, and May, W E (2001), CCQM-K6: Key Comparison on the determination of cholesterol in serum (Sevres, France: International Bureau of Weights and Measures).

# 6

## Measurement Uncertainty

### 6.1 Introduction

One of the great revolutions in metrology in chemistry has been the understanding of the need to quote an appropriate measurement uncertainty with a result. For some time, a standard deviation determined under not particularly well-defined conditions was considered a reasonable adjunct to a measurement result, and multiplying by the appropriate Student's $t$ value gave the 95% confidence interval. But knowing that in a long run of experiments repeated under identical conditions 95% of the 95% confidence intervals would include the population mean did not answer the fundamental question of how good the result was. This became evident as international trade burgeoned and more and more discrepancies in measurement results and disagreements between trading partners came to light. To determine if two measurements of ostensibly the same measurand on the same material give results that are equivalent, they must be traceable to the same metrological reference and have stated measurement uncertainties. How to achieve that comparability is the subject of this chapter and the next.

When making a chemical measurement by taking a certain amount of the test material, working it up in a form that can be analyzed, calibrating the instrument, and performing the measurement, analysts understand that there will be some doubt about the result. Contributions to uncertainty derive from each step in the analysis, and even from the basis on which the analysis is carried out. An uncertainty budget documents the history of the assessment of the measurement uncertainty of a result, and it is the outcome of the process of identifying and quantifying uncertainty. Although the client may only receive the fruits of this process as (value ± expanded uncertainty), accreditation to ISO/IEC 17025 requires the laboratory to document how the uncertainty is estimated.

Estimates of plutonium sources highlight the importance of uncertainty. The International Atomic Energy Agency (IAEA) estimates there are about 700 tonnes of plutonium in the world. The uncertainty of measurement of plutonium is of the order of 0.1%, so even if all the plutonium were in one place, when analyzed the uncertainty would be 700 kg (1000 kg = 1 tonne). Seven kilograms of plutonium makes a reasonable bomb.

## 6.2 Definitions of Measurement Uncertainty

The *International Vocabulary of Basic and General Terms in Metrology* (VIM), second edition (ISO 1993a), defines uncertainty of measurement as the "parameter, associated with the result of a measurement, that characterizes the dispersion of the values that could reasonably be attributed to the measurand" (term 3.9). In the third edition (Joint Committee for Guides in Metrology 2007), the term "measurement uncertainty" is defined (term 2.27), and "reasonably attributed" is replaced by "attributed to a measurand based on the information used." This definition emphasizes that the estimate of uncertainty is only as good as the information used to calculate it, so a report of an uncertainty should also state what information was used to calculate the uncertainty.

Measurement uncertainty characterizes the extent to which the unknown value of the measurand is known after measurement, taking account of the information given by the measurement. Compared with earlier, more complicated definitions, the current definition of measurement uncertainty is quite simple, although it does not give much guidance on how it is to be estimated. What information should be used, and how should one use it?

There are several terms used in measurement uncertainty that must be defined. An uncertainty arising from a particular source, expressed as a standard deviation, is known as the *standard measurement uncertainty* ($u$). When several of these are combined to give an overall uncertainty for a particular measurement result, the uncertainty is known as the *combined standard measurement uncertainty* ($u_c$), and when this figure is multiplied by a *coverage factor* ($k$) to give an interval containing a specified fraction of the distribution attributable to the measurand (e.g., 95%) it is called an *expanded measurement uncertainty* ($U$). I discuss these types of uncertainties later in the chapter.

The approach to measurement uncertainty that is espoused by many authorities, including the VIM, is embodied in a standard called the *Guide to the Expression of Uncertainty in Measurement*, universally referred to as

"GUM" (ISO 1993b). The GUM covers all measurement, and so for chemists a more pertinent guide has been published by EURACHEM (2000), which may be freely downloaded from the web (http://www.eurachem.ul.pt/). This guide, titled "Quantifying uncertainty in analytical measurement," is known by the acronym QUAM.

## 6.3  Where Does Uncertainty Come From?

There are two basic sources of uncertainty in a measurement result: definitional uncertainty and uncertainty in the measurement. Definitional uncertainty means that there will always be an inherent uncertainty in the way we describe what is being measured. For example, the measurement of "copper in lake water" may engender the question of whether total copper, or $Cu^{2+}$ (aq) is meant, or whether it is some average value that is envisaged or copper in water near the surface or in the sediment. However the definition might be refined, the analytical method might not give a result for that particular measurand. Organic analysis opens questions of isomeric forms, and the analysis of surface concentrations requires a definition of the surface (geometric, accessible by a particular molecule, and so on). As more information is given the measurement uncertainty also changes, usually for the better. Definitional uncertainty is considered to be the lower limit of measurement uncertainty: when measurements are maximally precise, definitional uncertainty accounts for the uncertainty. The first step in estimating measurement uncertainty is to define the measurand, and so it is here that definitional uncertainty arises. Uncertainty from the measurement itself is usually considered in terms of systematic and random elements, each of which requires a different approach to estimation. However, in different circumstances a random factor may become a systematic one and vice versa, and while keeping the notions of random and systematic errors, a holistic approach to measurement uncertainty is more satisfactory. Once the result has been corrected for any quantified and significant systematic errors (bias), some information used to estimate measurement uncertainty will be from repeated measurement results and will be treated statistically (Type A), and other information will be from different sources (Type B) that will be eventually combined with Type A information to give the overall estimate.

### 6.3.1 Error and Uncertainty

Modern measurement has tried to get away from the traditional concepts of accuracy or trueness. These concepts are based on the false assumption that there is a single true value that lurks in the measurement system and that in principle can be accessed by a sufficient number of measurements done with sufficient attention to detail. In reality, the measurement defines to a large extent what is measured, and the uncertainty can only describe a range in which a there is a reasonable chance of finding a value that might be properly attributed to the measurand. A true quantity value can only be defined as a quantity value consistent with the definition of a quantity, allowing many true values. The estimation of measurement uncertainty is therefore not an exercise in blame, but a process designed to increase understanding of the measurement result by assessing the factors that influence the measurement. It is convenient, because of the way the statistics are treated, to distinguish between random measurement errors, the standard deviation of which can be estimated from repeated measurements, and systematic measurement errors that are one-way deviations, which can be measured or otherwise estimated and then corrected for or included in the uncertainty budget. These errors arise from random and systematic effects, which can be identified, even if the resulting error cannot always be completely quantified.

There is one genuine error that can affect an analytical measurement. Absolute blunders happen from time to time, from human causes, catastrophic equipment failure, or acts of God. When the effect and cause are evident, the experiment can usually be repeated after corrective action is taken. The incident is not totally expunged from recorded history. A good quality system will require logging of such events, and some form of postmortem examination should be carried out. The analytical chemist must be ever vigilant for spurious errors that are not so clear. A bubble in a flow cell might not be visible but can cause erroneous results; transcription errors, by their nature, are probably not noticed at the time and can lead to all sorts of problems; and cross-contamination of test items can render any analysis worthless. In DNA analysis for forensic purposes, contamination of samples by the ubiquitous DNA of the investigating officers and analysts is a continuing specter. Such blunders are often identified as suspect outliers when results are being scrutinized. If enough measurements are made, statistical tests such as Grubbs's test (see chapters 2 and 4) can help. However, a result that is outside the 95% confidence interval of a mean is expected to occur 5 times out of 100 for measurements that are genuinely giving results that are part of the bona fide population. Culling extreme but genuine results truncates the distribution and tends to give a smaller dispersion than actually exists. When results are made in duplicate, a difference between the results that is greater than the repeatability limit ($2 \sqrt{2} \, s_r$) must be investigated and a third measurement taken. ISO 5725 (ISO 1994) gives guidance on the procedure. This is a sensible step-by-step procedure that leads to a major investigation if results are generated that are no longer in statistical control. It is considered unwise simply to exclude a result on a statistical basis alone, with some other evidence of a problem.

### 6.3.2  Bias and Recovery

Any result for known, quantified systematic effects must be corrected. How this is accomplished is discussed later. If the result is to be traceable to international standards, and not be an empirical result defined by the method used, systematic effects must be fully taken into account. Either they must be measured and the subsequent analytical results corrected, or they must be included in the measurement uncertainty. Measuring systematic effects and correcting the results will lead to greater understanding of the system and a lower measurement uncertainty, but there are times when it is not practical to estimate these effects, and then the measurement uncertainty is increased to allow for them. Some authorities, for example Codex Alimentarius, have decided that certain measurements made on foods should not be corrected, even when effects are known. The argument is that results, even if they are not metrologically traceable to international references, are comparable among the community that generate and use them. In this case fit-for-purpose overrides metrological purity.

### 6.3.3   Sources of Uncertainty

#### 6.3.3.1   Type A Uncertainty: Random Error

Type A uncertainty is derived from a statistical treatment of repeated measurement results. It is expressed as a standard deviation and combined with other uncertainty estimates to give the measurement uncertainty of the measurement result. Influence factors that impinge on the measurement will cause changes in the result from determination to determination. Because of its random nature, the distribution of the cumulative effect will tend to give more small deviations from the mean than large deviations from the mean, and it is symmetrical (i.e., a deviation will be just as likely to increase the result as to decrease it). Taking the average of a large number of measurement results is expected to reduce the effect of random factors through cancellation of positive and negative effects. Remember from chapter 2 that taking a large number of measurements does not reduce the standard deviation, it just gives a better estimate of the population standard deviation. The standard deviation of the mean is reduced, however, and does go to zero as the number of measurements averaged increases to infinity (through the $1/\sqrt{n}$ term). As many analytical measurements are made only in duplicate, it is more likely to have to estimate the standard deviation, perhaps from validation studies or other internal quality control experiments, and then include it in the measurement uncertainty, than to have the luxury of asserting that so many measurements have been done on the test samples that the standard deviation of the mean may be taken as zero.

#### 6.3.3.2   Type B Uncertainty:
#### Systematic Errors

Systematic effects are different from random ones. They are always in one direction and always (in their purest form) of the same magnitude. Averaging therefore does no good at all. The mean of a systematic error of magnitude $\Delta x$ is $\Delta x$, and so although repeated measurements may reduce the random effects sufficiently to make the systematic effects obvious, and therefore correctable, they cannot get rid of these effects. At best experiments can be performed to measure systematic errors which are then used to correct the initial measurement result. Otherwise the systematic errors are estimated, and they become a component of the overall measurement uncertainty. In the jargon of measurement uncertainty these components are known as Type B components. The methodology given later in the chapter is a recipe for identifying systematic effects, estimating their magnitude and combining them with standard deviations of random effects to give the overall measurement uncertainty.

Some examples include evaluation of uncertainty components associated with published values (i.e., the analyst did not measure them), uncertainties in a certificate of a certified reference material, manufacturer's statements about the accuracy of an instrument, or perhaps even personal experience. The latter could be viewed as an opportunity for anyone to just make up an uncertainty, but experience does count for something, and it is indeed usually better than nothing. Leaving out a component because of lack of exact knowledge immediately underestimates the uncertainty.

### 6.4  Using Measurement Uncertainty in the Field

#### 6.4.1  Measurement Uncertainty Requirements for Accreditation

Section 5.4 of the ISO/IEC standard 17025 (ISO/IEC 2005) requires "Testing laboratories shall have and shall apply a procedure to estimate the uncertainty of measurement," and in a test report "where applicable, a statement on the estimated uncertainty of measurement; information on uncertainty is needed in test reports when it is relevant to the validity or application of the test results, when a client's instruction so requires, or when the uncertainty affects compliance to specification limits" (ISO/IEC 2005, section 5.10). Although the reporting clause leaves open the possibility of not including the measurement uncertainty of a result, I believe that the added value to the client of a proper measurement uncertainty statement far outweighs any temporary problems that may be caused by unfamiliarity with the measurement uncertainty concept.

#### 6.4.2  Clients and Measurement Uncertainty

There are apocryphal stories of a statement of measurement uncertainty in a test report being used court and eliciting questions from defense counsel such as "So you admit that you really do not know what the amount of drugs my client was alleged to be in possession of?" followed by a rapid end of the case, not in the favor of the prosecution. The possibility of confusion, especially in the strange and arcane world of the legal system, is often raised, perhaps by laboratories that do not want to bother with an estimate of measurement uncertainty. There may also be a feeling that the professional status of an analyst is compromised by admitting that the result is open to doubt. This is a wrong view of our contribution to the world. It overestimates the importance of the analyst, implies that analysts have a very poor opinion about the capacity of our clients to understand rational argument, and demeans the worth of our product. Although the courts demand that scientific evidence be clear and unambiguous, they also have to deal with much evidence that is highly uncertain. Did the aging, myopic witness really see the defendant running from the scene of the crime? Is the implausible defense story just possibly true? In the context of a trial, often the scientific evidence is the most easy to assimilate, even with uncertainty of measurement.

Analytical chemistry has been helped in recent years by the ubiquity of DNA evidence. Here the measurement uncertainty is not the issue, but simply the probability that a particular electrophoretic pattern could come from someone other than the defendant. Statements such as "It is 1,000,000 times more likely that the DNA found at the crime scene came from Mr. X, than came from an unrelated, white male" are put to juries, and it has been shown that mostly, with good direction from the judge, the message has been understood. In the next section I explore the consequences of treating the quoted measurement result with its uncertainty as an interval containing the value of the measurand with a certain probability.

### 6.4.3  Measurement Uncertainty As a Probability

The standard uncertainty of a measurement result tells about the spread of results (dispersion) that would be expected given the basis on which it was estimated. It has the properties of a standard deviation, and with the appropriate degrees of freedom, a probability of finding a particular result can be calculated. The concept is illustrated in figure 6.1.

Suppose a very large number of measurements could be made under conditions of measurement that allow all possible variation that could occur, including systematic effects from calibrations of balances, glassware, and so on. Also suppose that the material being measured was identical for all of these measurements, which were correctly applied with any identified systematic effects corrected for. The reality of measurement uncertainty is that these measurements would not be identical but would scatter around the value of the measurand. In the absence of any other information, this dispersion is assumed to be normally distributed, which can be described by two parameters, the mean ($\mu$) and the standard deviation ($\sigma$). It is not

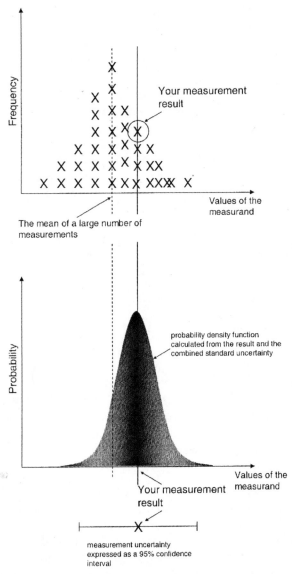

**Figure 6.1.** An illustration of the concept of dispersion of measurement results. Upper panel: results that might be obtained by multiple bona fide measurements. Each result is represented by a cross. Lower panel: The spread of results expressed as a normal probability density based on a single result and knowledge of the measurement uncertainty.

possible to conduct experiments like these. The analyst cannot do enough experiments, nor allow all factors to influence the result. It is possible, under repeatability and reproducibility conditions, to arrive at standard deviations that encompass a large part of the total dispersion referred to in the definition of measurement uncertainty, but the analyst will need to add estimates of systematic effects.

The EURACHEM (2000, p5) guide states that the expanded uncertainty "provides an interval within which the value of the measurand is believed to lie with a higher level of confidence." Usually the coverage factor of 2 implies this confidence is 95%. The GUM (ISO 1993b) is more circumspect and reminds us that the distributions must be normal and defines the expanded uncertainty (section 2.3.5) as giving "an interval that may be expected to encompass a large fraction of the distribution of values that could reasonably be attributed to the measurand." Whether the expanded uncertainty should be interpreted as a range in which about 95% of measurement results would fall if they were conducted under the referenced conditions or whether the analyst reports that the (true) value of the measurand lies in this range with 95% probability may depend on the flexibility of definitions. In practice the expanded uncertainty is used mostly in the latter sense, certainly when it comes to assessing results against each other or against regulatory or specification limits. This theme is expanded in section 6.6.4.8.

## 6.5 Approaches to Estimating Measurement Uncertainty

The late William Horwitz, who is the leading exponent of "top down" measurement uncertainty, offers the following four approaches to estimate measurement uncertainty (Horwitz 2003):

1. Calculate a confidence interval from the standard deviation of replicate measurements.
2. Use a "bottom-up" approach. Horwitz (2003) describes this approach as "recommended by the bible on uncertainty, rubber stamped by nine international organizations" (ISO 1993b).
3. Use a "top-down" approach using reproducibility and repeatability standard deviations from an interlaboratory study by the Harmonized IUPAC/AOAC protocol (Horwitz 1995) or ISO 5725 (ISO 1994).
4. Apply one of the formulas derived by Horwitz relating the relative standard deviation to concentration expressed as a mass fraction.

The tone of Horwitz's comments on the bottom-up approach (known as the GUM approach) imply that he does not approve. Later in the same paper he makes his objections clear: "This absurd and budget-busting approach (for analytical chemistry) arose from metrological chemists taking over in entirety the concepts developed by metrologists for physical processes measured with 5–9 significant figures (gravitational constant, speed of light, etc.) and applying them to analytical chemistry measurements with 2 or 3 significant figures" (Horwitz 2003). As a chemical metrologist I might not agree entirely with this assessment, but Horwitz does have a point that must be considered. Why bother going through a long, complicated, theoretically-based method when our own quality control and method validation data give as good an answer? Below, I consider each approach in turn.

### 6.5.1  Repeatability and Intralaboratory Precision

The repeatability ($s_r$) can be used to check duplicate repeats during normal operation of a method (see chapter 1). On its own, repeatability is not a complete basis for estimation of measurement uncertainty because it omits many effects that contribute to the bias of measurements made within a single laboratory. However, combined with a good estimate of the run bias, the intralaboratory precision, obtained from quality control data, can be used to give an estimate of measurement uncertainty. See section 6.6.3.2 for details on correction for bias and recovery.

### 6.5.2  Bottom-up or GUM Approach

Despite Horwitz's criticism of the bottom-up approach, it is still sanctioned by seven international organizations (Horwitz states that it is sanctioned by nine organizations, but I could only find seven: International Organization for Standardization, International Bureau for Weights and Measures, International Union of Pure and Applied Chemistry, International Union of Pure and Applied Physics, International Electrotechnical Commission, International Federation of Clinical Chemistry, and International Organization for Legal Metrology; he might have included CITAC and EURACHEM, who both espouse the approach). I describe the bottom-up approach in great detail in section 6.6.

### 6.5.3  Top-down from Interlaboratory Studies

The top-down approach has become the antithesis to the bottom-up GUM approach. Having results from a large number of laboratories, each using the prescribed method, but otherwise changing the analyst, time, location, and equipment gives the greatest possible opportunity for effects to be randomized and therefore contribute to the reproducibility. Can the interlaboratory reproducibility simply be asserted to be the measurement uncertainty of your result? Possibly, but not necessarily. Remember the measurement uncertainty is of a result, not a method, and so the uncertainty contributions that have been randomized in the reproducibility should be included, or excluded in the uncertainty of each particular result. Additional effects that might have to be considered are discussed below. There might also be effects that are randomized across the laboratories that do not apply to measurements made in a particular laboratory, and so the reproducibility can be an overestimate of the uncertainty.

The top-down approach is often used when there are method validation data from properly conducted interlaboratory studies, and when the laboratory using reproducibility as the measurement uncertainty can demonstrate that such data are applicable to its operations. Chapter 5 describes these types of studies in greater detail. In assigning the reproducibility standard deviation, $s_R$, to the measurement uncertainty from method validation of a standard method, it is assumed that usual laboratory variables (mass, volume, temperature, times, pH) are within normal limits (e.g., ± 2°C for temperature, ± 5% for timing of steps, ± 0.05 for pH). Clause 5.4.6.2 in ISO/IEC 17025 (ISO/IEC 2005) reads, "In those cases where a well-recognized test method specifies limits to the values of the major sources of uncertainty of measurement and specifies the form of presentation of the calculated results, the laboratory is considered to have satisfied this clause by following the test method and reporting instructions."

Interlaboratory studies usually provide the test material in a homogeneous form that can be subsampled without additional uncertainty. In the field, if the result is attributed to a bulk material from which the sample was drawn, then sampling uncertainty needs to be estimated. Again, proper definition of the measurand is important in understanding where to start adding in uncertainty components.

Another result of the requirements to distribute homogeneous test material is that some pretreatment steps usually performed on routine samples are not done on the interlaboratory sample. For example, in interlaboratory rounds for the determination of metals by inductively coupled plasma, the test material is often a solution of the metals in water. Often a method requires extraction of the metals by acid digestion of the field sample, a procedure that will have significant uncertainty in terms of recovery of analyte.

If there is an inherent bias in the method, usually estimated and reported in method validation studies, its uncertainty needs to be included. However, if run bias is estimated during the analysis, the method bias will be subsumed, and the local estimate of uncertainty of any bias correction should be used. This might have the effect of lowering the uncertainty because not all the laboratory biases of the participants in the study are now included.

If the method reproducibility is attributed to the measurement uncertainty, it has to be shown that the method was used exactly according to the published protocol. Any variation in the method or in the sample as defined in the scope of the method might add uncertainty and should be scrutinized.

So, although reproducibility values from method validation studies can be useful, they should not be used without thought. You might have a smaller uncertainty.

### 6.5.4  Horwitz Formulas

In a seminal study in 1980, Horwitz and colleagues compiled interlaboratory reproducibility values for a large number of chemical analyses (around 7500, later expanded to more than 10,000) and observed that there was a trend to increased reproducibility (expressed as a relative standard deviation, RSD) with smaller concentration, expressed as a mass fraction ($x$) (i.e., 1 mg kg$^{-1}$ = 1 ppm and $x = 10^{-6}$). An empirical fit of the compiled data gave the Horwitz relationship

$$R = 2^{(1-0.5\log_{10}x)} \tag{6.1}$$

where $R$ is the relative reproducibility standard deviation expressed as a percentage. Horwitz has since expressed this as

$$R = 0.02\ x^{-0.1505} \tag{6.2}$$

or as $R = \sigma_H/x$, where $\sigma_H$ is the "Horwitz reproducibility":

$$\sigma_H = 0.02x^{0.85} \tag{6.3}$$

Subsequent studies have largely confirmed the general trend, although at the extremes the equation has been amended (see figure 6.2). For mass fractions > 0.13, the reproducibility follows

$$\sigma_H = 0.01x^{0.5} \tag{6.4}$$

$$\text{or } R = x^{-0.5} \tag{6.5}$$

and for concentrations < 10 ppb, a constant RSD of between 1/5 and 1/3 expresses published results better. It is argued that an RSD >> 30 or 40% is unlikely to give fit for purpose results.

Some use the term "Horrat" for the ratio $s_R/\sigma_H$ and argue that a Horrat much greater than unity implies that something is wrong with the collaborative trial that produced $s_R$. Horwitz and colleagues (2001) surveyed a database of recoveries of pesticide in agricultural samples and showed that the average Horrat was 0.8.

## 6.6  Steps in Estimating Bottom-up Measurement Uncertainty

Performing an estimate of measurement uncertainty is a structured process. It need not take all day, but the effort can be considerable. One of my PhD

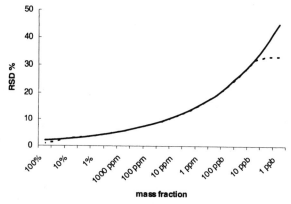

**Figure 6.2.** Horwitz horn curve of relative standard deviation as a function of mass fraction of analyte (*x*). The solid curve is $\log_2(R) = 1 - 0.5 \log_{10}(x)$ (Horwitz et al. 1980). Dotted lines at the extremes are modifications suggested after the original paper (Thompson 2004).

students spent three years validating quantitative nuclear magnetic resonance and providing an uncertainty budget. As with all aspects of quality assurance, the expenditure of time and money must be weighed against the expected benefit. For a routine method, for which there good validation data are available, measurement uncertainty sufficient to satisfy the requirements of accreditation to ISO/IEC 17025 can be estimated within a reasonable time. Even in the case of a quick (but not necessarily dirty) estimate, I recommend going through the five steps to uncertainty enlightenment:

1. Specify the measurand.
2. Identify major high-level sources of uncertainty.
3. Quantify uncertainty components.
4. Combine significant uncertainty components.
5. Review estimates and report measurement uncertainty.

Each step can be done in a short time, but I strongly recommend that they are all done. In doing so, a lot will be learned about the analysis, even if the analyst has been using the same method for years. The above steps are similar to those recommended by EURACHEM and in the GUM, except I have qualified the sources in step 2. Choose the most obvious source (possibly the reproducibility) and then add other factors as necessary. The academics and national measurement institutes will still delight in providing a full GUM-blessed approach, and this is entirely appropriate for some measurements, but for the majority of readers of this book, I hope my method will lead to sensible estimates that will satisfy clients, accreditation bodies, and the consciences of laboratory managers. This method recognizes Horwitz's problems with GUM, while keeping the approach of understanding the method and identifying sources of uncertainty, which, I believe, has merit.

### 6.6.1  Specify the Measurand

Stating exactly what will be measured helps define the scope of the uncertainty budget and flags any definitional uncertainty that may need to be accounted for. For example, there is a huge difference between "copper in a sample of water from Lake Eyre" and "copper in Lake Eyre." The latter has to take account of the enormous uncertainties in obtaining a representative sample, or even defining what is meant by "in Lake Eyre." Measurement uncertainty is often denigrated by environmental chemists because invariably the uncertainty of sampling, which is carefully taken into account before any conclusions are made, is considerably greater than the uncertainty of measurement. So why bother about measurement uncertainty? The answer is that if, after due consideration, the conclusion is that measurement uncertainty is insignificant compared with sampling, then for this particular problem measurement uncertainty may be ignored in the overall estimate of the uncertainty of the result after duly recording and justifying this action. Reporting a pH measurement of a given sample as 6.7 ± 0.0 is possible. This does not mean there is no measurement uncertainty, but at the quoted level of precision the measurement uncertainty is zero. In the context of a sampling program of some effluent water, for example, the pH might eventually be reported as 6 ± 2. Knowledge of the measurement uncertainty might also lead the analyst to choose a cheaper analytical method with greater, but still fit-for-purpose measurement uncertainty.

### 6.6.2  Identify Sources of Uncertainty

Specifying the measurand implies that the measurement method and relevant equations are specified. This provides a template for examining sources of Type B uncertainties. For example, a simple titration from which the concentration of an acid is to be measured by the formula

$$M_2 = \frac{M_1 V_1}{V_2} \tag{6.6}$$

where $M_2$ is the desired concentration, $M_1$ is the concentration of the standard alkali, $V_1$ is the end point volume of the titration, and $V_2$ is the volume of the acid solution pipetted into the reaction flask, gives three independent sources of errors. If the measurand was, for example, the percent purity expressed as a mass fraction of a sample of potassium hydrogen phthalate, then equations representing this quantity could be combined with the subsequent operations that determined the other quantities involved.

$$P = \frac{m_{\text{analyzed}}}{m_{\text{weighed}}} \times 100\%$$

$$= \frac{M_2 V_{\text{analyzed}}}{W_{\text{PHP}} m_{\text{weighed}}} \times 100\%$$

$$= \frac{M_1 V_1 V_{\text{analyzed}}}{V_2 W_{\text{PHP}} m_{\text{weighed}}} \times 100\% \qquad (6.7)$$

In the grand equation for the purity $(P)$, $V_{\text{analyzed}}$ is the volume of solution in which the mass, $m_{\text{weighed}}$, of the sample of potassium hydrogen phthalate was dissolved, $W_{\text{PHP}}$ is the molar mass of potassium hydrogen phthalate, and the other terms are as defined for equation 6.6. Now there are six quantities that can be assessed for uncertainty, and perhaps this could be expanded further—for example, if the standard alkali were made up from solid of known purity. One way to visualize the relationships between these factors is a cause-and-effect diagram, also known as an Ishikawa diagram, after the person who popularized its use. The Ishikawa diagram was introduced in a general way for describing processes in chapter 4, and here it is applied specifically to uncertainty components.

### 6.6.2.1  Cause-and-Effect Diagrams in Measurement Uncertainty

To construct a cause-and-effect diagram of uncertainty sources from the information contained in the procedures and equations of an analytical method, follow these steps. First, draw a horizontal right-facing arrow in the middle of a sheet of paper. Label the arrow end with the symbol for the measurand. Starting from the sources identified by the equation for the value of the measurand, draw arrows to this line at about 45°, one for each of the quantities in your equation plus any other sources identified that are not already counted, plus one for repeatability. Label the start of each arrow with a symbol for the quantity. Figure 6.3 shows a draft cause-and-effect diagram for the purity of the acid.

These diagrams are very helpful because each side arrow may be treated as a problem in its own right, and so can be embellished with arrows to stand for components of its uncertainty, and so on. Some possible candidates are added to the basic diagram in figure 6.4. Even at this stage I advocate leaving out obvious components that you judge to be insignificant. Molar masses are known to better than 0.1% (a couple of orders of magnitude better in many cases), so they will rarely impinge on deliberations of measurement uncertainty of an analytical measurement. The mass of the test portion of potassium hydrogen phthalate taken for analysis has also been considered to have negligible uncertainty. This decision requires more careful justification, but in this case about 1.0 g was weighed on a balance to ± 0.1 mg, so its uncertainty was deemed small compared to other effects.

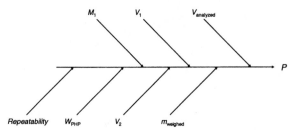

**Figure 6.3.** First draft of a cause-and-effect diagram for the measurement of the purity of a sample of potassium hydrogen phthalate. See text for description of symbols.

It is important not to assume that this or that effect always has negligible uncertainty; each factor must be assessed for a particular result. For example, in the quantitative NMR work in my laboratory, we could just see the effect of the uncertainty of molar masses. As a rule of thumb, if the uncertainty component is less than one-fifth of the total, then that component can be omitted. This works for two reasons. First, there are usually clear major components, and so the dangers of ignoring peripheral effects is not great, and second, components are combined as squares, so one that is about 20% of the total actually contributes only 4% of the overall uncertainty. There is a sort of chicken-and-egg dilemma here. How do you know that a component is one-fifth of the total without estimating it and without estimating the total? And if you estimate the contribution of a factor to uncertainty,

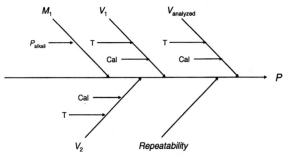

**Figure 6.4.** Second round of a cause-and-effect diagram for the measurement of the purity of a sample of potassium hydrogen phthalate. $T$ is temperature, and Cal is the calibration of the volume. The molar mass of potassium hydrogen phthalate and the mass of potassium hydrogen phthalate dissolved for analysis are excluded from the diagram because they have negligible uncertainties.

then there is no point in leaving it out, the hard work having been done. The answer is that the five-step process is iterative, and an experienced analyst will have a feel for the orders of magnitude of uncertainties of common contributors. There must always be a final sanity check of an uncertainty budget, and depending on how important it is that the estimate be as complete as possible, some components might be thoroughly evaluated before, in hindsight, you decide they could have been dispensed with.

### 6.6.2.2  Typical Uncertainty Components

Weighing on an appropriate balance is usually a very accurate procedure, and therefore is often a candidate for the insignificance test. Uncertainty components of weighing on a modern electronic balance are given in table 6.1.

Volume measurements include the use of pipettes, volumetric flasks, measuring cylinders, and burettes. Each has different calibration uncertainties that are detailed in the manufacturer's information supplied with the

Table 6.1.  Uncertainty components of weighing

| Component | Calculation | Typical values for a 4-decimal place tared balance |
|---|---|---|
| Repeatability | Obtain as standard deviation of repeated weighings or assume part of repeatability of measurement; $u = s$. | $u = 0.2$ mg |
| Linearity of calibration | Main source of uncertainty arising from the balance itself; take manufacturers figure ($\pm a$ mg) and treat as rectangular distribution, $u = a/\sqrt{3}$ | $u = 0.15$ mg |
| Systematic error of balance (also known as sensitivity) | For tared balances, or weighing by difference, as long as the masses are not too different any systematic effects cancel | 0 mg |
| Readability of scale | Value of smallest digit | 0.1 mg |
| Buoyancy correction | Conventional masses are based on weighing at sea level with air density 1.2 kg m$^{-3}$ and sample density 8000 kg m$^{-3}$ Rarely needed. | Usually ignore with error in true mass of less than 0.01 % |
| Moisture uptake by sample | Try to avoid, but be aware hygroscopic material could absorb water in a humid atmosphere leading to significant changes in mass | Change conditions (weigh in dry atmosphere) to cause uncertainty to be negligible |

glassware. Most countries have minimum standards for tolerances of calibrated glassware. Temperature changes during a series of measurements will lead to uncertainty in the result, as will the analyst's ability to fill to the mark and correctly deliver the desired volume (see figure 6.5). These different components are brought together in table 6.2.

In addition to the above sources that should be assessed, for burettes used in titration there is the end point error. There is the repeatability of the end point determination, which is in addition to the repeatability of reading the burette scale, but is part of the repeatability calculated from a number of replicate, independent analyses. There is also uncertainty about the coincidence of the end point, when the color of the solution changes, and the equivalence point of the titration, when a stoichiometric amount of titrant has been added.

Although changes in temperature, whether of the laboratory or the oven, the injection port of a gas chromatography instrument, or any other con-

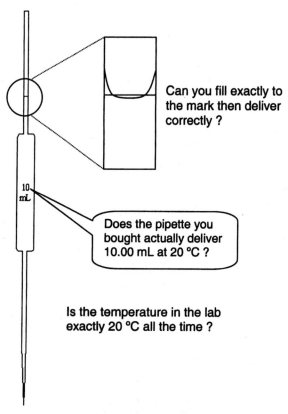

**Figure 6.5.** Uncertainties in delivering 10 mL by a pipette. (Based on figure 1.1 in Hibbert and Gooding 2005.)

Table 6.2. Uncertainty components of volume measurements

| Component | Calculation | Typical values for a 100-mL volumetric flask |
|---|---|---|
| Repeatability | Obtain as standard deviation of repeated weighings or assume part of repeatability of measurement (this may be subsumed in the overall repeatability) | $u = 0.02$ mL |
| Calibration | Take from manufacturers specification ($\pm a$ mg) and treat as triangular distribution, $u = a/\sqrt{6}$ | $u = 0.04$ mL |
| Temperature | The volume expansion of water is $0.00021°C^{-1}$; therefore if the uncertainty of temperature is $u_T$, then $u = 0.00021 \times V \times u_T$ | $u = 0.04$ mL for a range of temperature of $\pm 4$ °C with a rectangular distribution $u_T = 4/\sqrt{3}$ |

trolled space, are sources of uncertainty, temperature rarely enters the measurement function directly. Temperature usually is found as an arrow-on-an-arrow in the cause-and-effect diagram, and its effect is calculated as a part of the uncertainty for the quantity being studied. Volumes of liquids and gases are sensitive to temperature, and variations in temperature should be accounted for in all volume calculations.

A certified reference material (CRM) used for calibration and to establish metrological traceability comes with a certificate that includes the uncertainty of the quantity value that is being certified. If this is quoted as a confidence interval, the coverage factor, $k$ (see below), will also be stated. In most cases $k$ is 2, giving a confidence interval covering about 95% of the distribution that can be attributed to the measurand. The standard uncertainty that must be used in the uncertainty calculations is determined by dividing the quoted half range (value ± half range) by $k$. In general, for a CRM value quoted as $x \pm U$ with coverage factor $k$, then

$$u = U/k \qquad (6.8)$$

If the CRM is used directly, then $u$ is all that is required. If, for example, the material is a pure reference material with certified purity that is to be dissolved in a buffer to provide a calibration solution, then the uncertainties of the dissolution step and the volume presented to the measuring instrument must also be included.

If a calibration function is used with coefficients obtained by fitting the response of an instrument to the model in known concentrations of calibration standards, then the uncertainty of this procedure must be taken into account. A classical least squares linear regression, the default regression in most spreadsheets and calculators, has three main assumptions: the linear model is valid, the response of the instrument is a random variable with constant variance (homoscedacity), and the independent variable (the concentration or amount of calibrant) is known without uncertainty. The model in terms of an equation is

$$Y = a + bx + \varepsilon \tag{6.9}$$

where $a$ is the intercept, $b$ is the slope of the linear relation, and $\varepsilon$ is a normally distributed random variable with mean zero and variance the repeatability variance of the measurement. You must be sure that these assumptions hold. In particular, the constancy of variance is rarely held over wide concentration ranges. More often the RSD is approximately constant, which leads to an increasing standard deviation with $Y$, in which case weighted regression should be used. When the calibration relation (equation 6.9) is inverted to give the concentration of an unknown ($\hat{x}_0$) from an observed response ($y_0$) the standard deviation of the estimate ($s_{\hat{x}_0}$) is given by

$$s_{\hat{x}_0} = \frac{s_{y/x}}{b} \sqrt{\frac{1}{m} + \frac{1}{n} + \frac{(y_0 - \bar{y})^2}{b^2 \sum_i (x_i - \bar{x})^2}} \tag{6.10}$$

where $m$ is the number of replicate observations of the response (mean $y_0$), $n$ is the number of points in the calibration line, $b$ is the slope of the calibration line, $\bar{y}$ and $\bar{x}$ are the means of the calibration data, and $x_i$ is the $i$th $x$ value from the calibration set. In equation 6.10, the term $s_{y/x}/(b\sqrt{m})$ represents the repeatability of the instrumental response to the sample. If an independent estimate of the repeatability is known, it can be used instead of $s_{y/x}$.

If there is any doubt about whether 100% of the analyte is presented to the measuring system or that the response of the calibrated system leads to no bias , then the assumptions must be tested during method validation and appropriate actions taken. If a series of measurements of a CRM (not used for calibration) leads to the conclusion that there is significant bias in the observed measurement result, the result should be corrected, and the measurement uncertainty should include the uncertainty of the measurement of the bias. If the bias is considered insignificant, no correction need be made, but measuring the bias and concluding that it is zero adds uncertainty (perhaps the bias was not really zero but is less than the uncertainty of its measurement). One approach to measurement uncertainty is therefore to include CRMs in the batch to be used to correct for bias, and then the uncertainty of estimation of bias, which includes the uncertainty of the quantity value of the CRM, is combined with the within-laboratory reproducibility. In some fields of analysis it is held that routine measurement and correction for bias is not possible, but it is recommended that an estimate of the bias be included in the measurement uncertainty. Although this is does not comply with international guidelines (ISO GUM, EURACHEM QUAM), which all recommend correction for bias, some strategies for including estimate of bias are given in the next section.

Recovery can also be a problem if the test portion of a sample must be prepared by derivatization, extraction and other chemical or physical steps. Even if the recovery can be shown to be 100%, there may still be effects that have uncertainty components arising from the preparation. An example is leaching a product for heavy metal analysis of the leachate. The amount that is extracted from the product depends on the leaching solution (kind of chemical, concentration), the temperature of leaching, and how long the product is leached. Uncertainties in each of these influence quantities can lead to variation in the measurement result. The effects should be quantified in the method validation studies, and their effects can be added as factors in the cause-and-effect diagram. For example, a simplified diagram for the determination of mercury in fish might look like that of figure 6.6.

A mass, $m_{fish}$, is weighed, digested in hot acid for 30 minutes, and then the solution is made up to a volume, $V$. This solution is analyzed by hydride generation atomic absorption (AA). The equation describing the concentration of mercury in a fish can be written

$$c_{fish} = \frac{c_{extract}V}{m_{fish}} = \frac{c_{digest}V}{m_{fish}} f_{time} f_{temp} f_{conc} \qquad (6.11)$$

where the concentration of the digested solution as measured by AA might not be the same as the extractable concentration that will give the appropriate concentration, $c_{fish}$. The effects that could lead to this inequality ($c_{extract} \neq c_{digest}$) might arise from the time and temperature of the digestion and the concentration of acid. Even if the factors $f_{temp}$, $f_{time}$, and $f_{conc}$ are all unity,

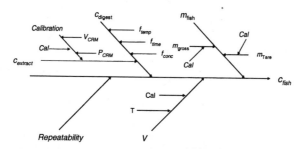

**Figure 6.6.** Cause-and-effect diagram for the analysis of mercury in fish showing the introduction of factors associated with the digestion process.

there still might be an uncertainty associated with the variability in these effects that needs to be accounted for.

If the measurand is an amount of a particular species in a bulk material that must be sampled before analysis, the heterogeneity of the sample must be considered. Examples are found in the environment, particularly solid sampling, in biological sampling where natural variation must be accounted for, and in process control of items that tend to have a large variability. The variance caused by sampling is estimated by measuring multiple samples analyzed independently. Again, sampling is rarely included in interlaboratory method validation studies and so this component will need to be combined with a reproducibility, if used.

### 6.6.3  Quantify Uncertainty Components

I have already suggested that the analyst might have some idea about the magnitude of the uncertainties as the cause-and-effect diagram evolves. Obviously, minor components such as the uncertainties of molar masses can usually be assessed and omitted at this early stage. The analyst should have a feel for typical uncertainties of masses, volumes, and temperatures, some of which are discussed below.

#### 6.6.3.1  Example of Volume Measurement

It is worthwhile to discuss the components of the standard uncertainty of a volume measurement here. The repeatability may be independently assessed by a series of fill-and-weigh experiments with water at a controlled temperature (and therefore density) using a balance so that the uncertainty of weighing is small compared with the variation in volume. Although this may be instructive, if the whole analysis is repeated, say, ten times, then the repeatability of the use of the pipette, or any other volume measurement is part of the repeatability of the overall measurement. This shows the benefit, in terms of reaching the final estimate of measurement uncertainty more quickly, of lumping together uncertainty components.

The manufacturer's calibration information must now be accounted for. You might be lucky and have a 10-mL pipette that is indeed 10.00 mL, but the manufacturer is only guarantees the pipette to be not less than 9.98 mL and not more than 10.02 mL. As the pipette is used throughout many experiments, then this possible systematic effect will not dissipate with repeated measurements and must be accounted for. There are two alternatives. The first is to calibrate the pipette in the laboratory. The fill-and-weigh experiments that would give the standard deviation of the measurement will also, from the mean, give an estimate of the volume of the pipette (see spreadsheet 6.1).

Suppose with 10 fill-and-weigh experiments, the mean volume of a pipette is 10.0131 mL with a standard deviation of $s = 0.0019$ mL. The standard deviation of the mean is $0.0019/\sqrt{10} = 0.0006$ mL. In future use of this

|    | A | B | C |
|----|---|---|---|
| 5 |  |  |  |
| 6 | Experiment | Volume /mL |  |
| 7 | 1 | 10.0104 |  |
| 8 | 2 | 10.0110 |  |
| 9 | 3 | 10.0116 |  |
| 10 | 4 | 10.0124 |  |
| 11 | 5 | 10.0129 |  |
| 12 | 6 | 10.0132 |  |
| 13 | 7 | 10.0139 |  |
| 14 | 8 | 10.0145 |  |
| 15 | 9 | 10.0150 |  |
| 16 | 10 | 10.0164 |  |
| 17 |  |  |  |
| 18 | mean | 10.0131 mL |  |
| 19 | s | 0.0019 mL |  |
| 20 |  |  |  |
| 21 | s(mean) | 0.0006 mL |  |
| 22 |  |  |  |

=AVERAGE(B7:B16)

=STDEV(B7:B16)

=B19/SQRT(10)

**Spreadsheet 6.1.** Data from 10 fill-and-weigh experiments of a 10-mL pipette. The masses have been converted to volumes, and the uncertainty calculation assumes the components of weighing and the volume calculation are negligible.

pipette, the volume used in the calculations is 10.013 mL and the uncertainty component is $u = 0.0006$ mL. Note that had this not been done, 10.0000 mL would be taken as the volume, and the uncertainty component would have been $0.02/\sqrt{6} = 0.0082$ mL (assuming a triangular distribution with $a = 0.02$ mL). Therefore the calibration has given a better estimate of the volume of the pipette with a smaller standard uncertainty. (Note: for these calculations it is better to perform them in a spreadsheet, which will maintain full precision until the last rounding to an appropriate number of significant figures. It is dangerous to keep truncating and rounding during the calculations.) A second way to account for the calibration of the pipette is to use a different pipette, chosen at random from a suitably large pool, every time the experiment is repeated. Now the standard deviation of the overall measurement, which already includes the variation from the use of a pipette, will be augmented by the pipette-to-pipette variation. There is no longer any need to include a specific component for the calibration. This discussion shows why the line between systematic and random effects can be moved by choice of the experiments.

The temperature effect can also be manipulated. If the temperature is measured at the time the experiment is performed, then the volume of the glassware used can be corrected for the difference between the nominal temperature of calibration of the glassware (usually 20°C). For example, suppose the mean temperature of the laboratory during an experiment is measured as 22.3°C, with a standard uncertainty of 0.5°C. The liquid in the pipette will expand (the glass expansion can be ignored), and so the delivery will be less than it is when performed at 20°C. The uncertainty of the temperature ($u = 0.5$°C) will include the uncertainties of the temperature measurement (reading thermometer, calibration of the thermometer) and the standard deviation of the mean of the temperature. The volume correction of a 10.013 mL pipette because of the temperature is $-0.00021 \times 2.3 \times 10.013 = -0.0048$ mL, with a standard uncertainty of $0.00021 \times 0.5 \times 10.013 = 0.0011$ mL. So now the volume is corrected again to $10.013 - 0.0048 = 10.0082$ mL, and the uncertainty component of temperature is 0.0011 mL, replacing 0.0042 mL, which would have been calculated from the estimated temperature variation in the laboratory of $\pm 4$°C (95% confidence interval). Again, an estimated systematic effect that is treated as a standard uncertainty has been turned into a genuine systematic effect that is measured and accounted for, with the uncertainty of that measurement result now being combined into the standard uncertainty of the overall measurement result. Spreadsheet 6.2 and figure 6.7 show the results of these measurements. As the estimate of the volume is improved, the uncertainty is also reduced. (It would not be sensible to make a measurement of a correction factor that had a greater uncertainty of the result than the Type B uncertainty.)

Is it all worth it? Calibration of glassware and careful temperature measurements during an experiment cause significant reduction in the uncertainty of the volumes of glassware. However, if the uncertainty in the glassware is only a minor contributor to the whole, you must judge each case on its merits and act accordingly.

### 6.6.3.2  Estimating Bias and Recovery

Systematic effects are estimated by repeated measurements of a CRM, suitably matrix matched. Any difference between the CRM and a routine sample for which the measurement uncertainty is being estimated should be considered and an appropriate uncertainty component added. Suppose a concentration measurement is routinely made in a laboratory that includes measurement of a CRM in the same run as calibration standards and unknowns. The bias ($\delta$) is given by

$$\delta = \bar{C}_{CRM} \text{ (measured)} - C_{CRM} \text{ (certified)} \tag{6.12}$$

where $C_{CRM}$(certified) is the certified concentration of the CRM, and $\bar{C}_{CRM}$(measured) is the mean measurement result for the CRM of $p$ replicates. The uncertainty of the measurement of the bias ($u_{bias}$) is

$$u_{bias} = \sqrt{\frac{s_r^2}{p} + u_{CRM}^2} \tag{6.13}$$

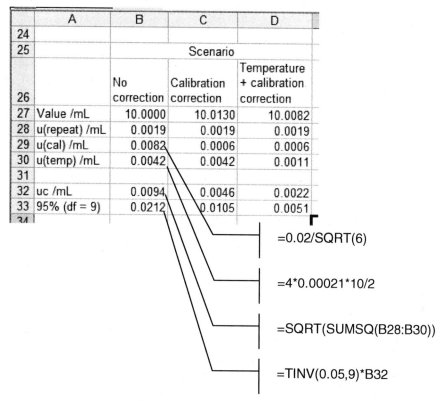

| | A | B | C | D |
|---|---|---|---|---|
| 24 | | | | |
| 25 | | | Scenario | |
| 26 | | No correction | Calibration correction | Temperature + calibration correction |
| 27 | Value /mL | 10.0000 | 10.0130 | 10.0082 |
| 28 | u(repeat) /mL | 0.0019 | 0.0019 | 0.0019 |
| 29 | u(cal) /mL | 0.0082 | 0.0006 | 0.0006 |
| 30 | u(temp) /mL | 0.0042 | 0.0042 | 0.0011 |
| 31 | | | | |
| 32 | uc /mL | 0.0094 | 0.0046 | 0.0022 |
| 33 | 95% (df = 9) | 0.0212 | 0.0105 | 0.0051 |
| 34 | | | | |

=0.02/SQRT(6)

=4*0.00021*10/2

=SQRT(SUMSQ(B28:B30))

=TINV(0.05,9)*B32

**Spreadsheet 6.2.** Calculation of the volume and measurement uncertainty of the delivery of a nominal 10-mL pipette under the scenarios given. These are graphed in figure 6.7.

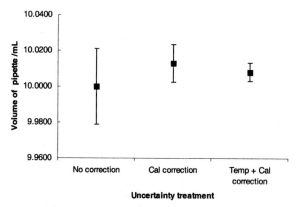

**Figure 6.7.** Value and 95% confidence interval on the volume delivered by a nominal 10.0000-mL pipette, with different corrections applied. Cal.: volume calibrated by 10 fill-and-weigh experiments. Temp: volume corrected for temperature measured in the laboratory.

where $s_r$ is the repeatability of the bias measurement, and $u_{CRM}$ is the uncertainty of the concentration of the CRM. The significance of the bias is tested by a one-tailed $t$ test at 95% confidence.

$$\delta > t_{0.05',p-1} \times u_{bias} \tag{6.14}$$

The test result for a sample $(C_{sample})$ measured $n$ times with mean $\bar{C}$ is then

$$C_{sample} = \bar{C} + \delta_{run} \tag{6.15}$$

with uncertainty

$$u_{sample} = \sqrt{\frac{s_r^2}{n} + u_{bias}^2} \tag{6.16}$$

The term $u_{bias}$ must be included even if the bias is considered insignificant. The expanded uncertainty is obtained by multiplying $u_{sample}$ by an appropriate coverage factor. Note that $s_r$ should be obtained from a suitably large (at least 10) number of repeats taken over a few days and batches of samples. A similar approach is taken for recovery, defined as

$$R = \frac{\bar{C}_{CRM}(\text{measured})}{C_{CRM}(\text{certified})} \tag{6.17}$$

and uncertainty

$$u_R = R\sqrt{\frac{1}{p}\left(\frac{s_r}{\bar{C}_{CRM}(\text{measured})}\right)^2 + \left(\frac{u_{CRM}}{C_{CRM}(\text{certified})}\right)^2} \tag{6.18}$$

### 6.6.4  Combine Significant Uncertainty Components

Once the uncertainty components have been identified and quantified as standard uncertainties, the remainder of the procedure to estimate uncertainty is a somewhat complicated but mostly straightforward. Most software products on the market will perform this task. Otherwise, some spreadsheet manipulation or mathematics must be done to reach the uncertainty. The combined standard uncertainty of a result is obtained by mathematical manipulation of the standard uncertainties as part of the uncertainty budget. These standard uncertainties may also be combinations of other uncertainties, and so on, as the branches and sub-branches of the cause-and-effect diagram are worked through. A combined standard uncertainty of a quantity $y$ is written $u_c(y)$.

### 6.6.4.1  The Mathematical Basis of
#### Combining Standard Uncertainties

The GUM approach described here has the advantage that each uncertainty component is designed to have the properties of a standard deviation, and so the rules for combining standard deviations of the normal distribution can be followed. The complete equation will be given, but it may be simplified to useable equations for the majority of applications.

For a general relation between the measurement result ($y$) and a series of input quantities ($x = x_1, x_2, \dots x_n$)

$$y = f(x) \tag{6.19}$$

the variance in $y$, $\sigma^2(y)$, is given by

$$\sigma^2(y) = \sum_{i=1}^{i=n} \left( \frac{\partial y}{\partial x_i} \sigma(x_i) \right)^2 + 2 \sum_{k=1}^{k=n-1} \sum_{j=k+1}^{j=n} \left( \frac{\partial y}{\partial x_k} \frac{\partial y}{\partial x_j} \mathrm{cov}(x_k, x_j) \right) \tag{6.20}$$

where $\mathrm{cov}(x_k, x_j)$ is the covariance between $x_k$, and $x_j$. If the square of the combined uncertainty is equated to the variance, then

$$u_c^2(y) = \sum_{i=1}^{i=n} \left( \frac{\partial y}{\partial x_i} u(x_i) \right)^2 + 2 \sum_{k=1}^{k=n-1} \sum_{j=k+1}^{j=n} \left( \frac{\partial y}{\partial x_k} \frac{\partial y}{\partial x_j} \mathrm{cov}(x_k, x_j) \right) \tag{6.21}$$

Equation 6.21 may be written in terms of the correlation coefficient between $x_k$, and $x_j$, $r_{k,j}$.

$$u_c^2(y) = \sum_{i=1}^{i=n} \left( \frac{\partial y}{\partial x_i} u(x_i) \right)^2 + 2 \sum_{k=1}^{k=n-1} \sum_{j=k+1}^{j=n} \left( \frac{\partial y}{\partial x_k} \frac{\partial y}{\partial x_j} u(x_k) u(x_j) r_{k,j} \right) \tag{6.22}$$

The differentials are partial, so when the differentiation is done with respect to one quantity, all the others are considered constant. The variable $\partial y/\partial x_i$ is known as the *sensitivity coefficient* (sometimes written $c_i$) with respect to $x_i$, and it tells how $y$ changes as $x_i$ changes. For the most simple case where $y = constant \times x$, the sensitivity coefficient is just the constant.

If all the input quantities are independent of each other (e.g., any variation in the mass of a test item has nothing to do with variations in the volume of solution that is introduced into the gas chromatograph), then $r$ is zero and equation 6.22 simplifies to

$$u_c^2(y) = \sum_{i=1,n} \left( \frac{\partial y}{\partial x_i} u(x_i) \right)^2 \tag{6.23}$$

and if $y$ depends on only a single quantity [i.e., $y = f(x)$], the equation becomes simpler still

$$u_c(y) = \frac{\partial y}{\partial x} u(x) \tag{6.24}$$

### 6.6.4.2 Added Components

For independent quantities of the same kind, expressed in the same units, the variances (the squares of the standard deviations) are multiplied by the sensitivity coefficient and summed to give the combined variance. Weighing by difference or calculating a titration volume by subtracting an initial reading from a final reading are examples. In these kinds of difference measurements, in which the measurements are made using the same instrument (balance, burette), constant systematic errors cancel, leaving only random errors or proportional systematic errors. Consider a delivered volume by difference expressed in liters where the starting and ending volumes are read in milliliters:

$$\Delta V = 0.001 \times (V_{end} - V_{start}) \tag{6.25}$$

where

$$c_{V_{end}} = \left(\frac{\partial \Delta V}{\partial V_{end}}\right)_{V_{start}} = 0.001 \text{ and } c_{V_{start}} = \left(\frac{\partial \Delta V}{\partial V_{start}}\right)_{V_{end}} = -0.001$$

then

$$u_c(\Delta V) = \sqrt{\left(0.001 \times u(V_{end})\right)^2 + \left(-0.001 \times u(V_{start})\right)^2} = 0.001 \times \sqrt{u(V_{end})^2 + u(V_{start})^2} \tag{6.26}$$

The mathematics are even simpler when contributions to the uncertainty of a single quantity are combined. Here the sensitivity coefficient is 1, and the individual uncertainties are just squared and summed. For example, for the combination of the standard uncertainties of the effects on the volume delivered by a pipette discussed above, which are repeatability, calibration uncertainty, and the effect of temperature, the square of the combined uncertainty is simply the sum of the squares of each effect:

$$u^2_c(V) = u^2_r(V) + u^2_{cal}(V) + u^2_{temp}(V) \tag{6.27}$$

therefore the combined uncertainty is

$$u_c(V) = \sqrt{u^2_r(V) + u^2_{cal}(V) + u^2_{temp}(V)} \tag{6.28}$$

Remember that equation 6.28 does not mean that a bunch of standard uncertainties can simply be averaged. The squaring and adding and square rooting are common to all these manipulations. Do not forget the old adage that you cannot add apples and oranges (except as pieces of fruit), so the uncertainty from variations in temperature, for example, must have been worked through to its effect on the volume before it can be used in equation 6.28. The temperature effect is turned into an uncertainty in volume by the relation between them [$u_{temp}(V) = 0.00021 \times V \times u_{temp}$]. The coefficient of expansion of water is $0.00021°C^{-1}$, and this multiplied by a volume in liters and uncertainty in the temperature in degrees Celsius leads to an uncertainty in volume in liters.

Uncertainties always combine according to equations like 6.28 even if the quantities are subtracted (i.e., the squares of the uncertainties are always added). This is why a calculation that eventuates in the subtraction of two large numbers of nearly equal magnitude (e.g., weighing the captain of the ship by weighing the ship with the captain on board and subtracting the mass of the ship when the captain is not on board), is notoriously beset by large uncertainty.

### 6.6.4.3 Quantities Multiplied or Divided

Measurement is the comparison of a known quantity with an unknown quantity. This comparison is often one of taking the ratio of indications of a measuring instrument. For the most simple case, if $y = x_2/x_1$ application of equation 6.23 gives

$$u_c^2(y) = \left(\frac{\partial y}{\partial x_1}\right)_{x_2}^2 u_c^2(x_1) + \left(\frac{\partial y}{\partial x_2}\right)_{x_1}^2 u_c^2(x_2) \qquad (6.29)$$

$$\left(\frac{\partial y}{\partial x_1}\right)_{x_2} = \frac{x_2}{x_1^2} = -\frac{y}{x_1}, \quad \left(\frac{\partial y}{\partial x_2}\right)_{x_1} = \frac{1}{x_1}$$

$$u_c^2(y) = y^2 \frac{u_c^2(x_1)}{x_1^2} + \frac{1}{x_1^2} u_c^2(x_2) = y^2 \frac{u_c^2(x_1)}{x_1^2} + \frac{x_2^2}{x_1^2} \frac{u_c^2(x_2)}{x_2^2} = y^2 \frac{u_c^2(x_1)}{x_1^2} + y^2 \frac{u_c^2(x_2)}{x_2^2} \qquad (6.30)$$

and therefore

$$\frac{u_c^2(y)}{y^2} = \frac{u_c^2(x_1)}{x_1^2} + \frac{u_c^2(x_2)}{x_2^2}$$

$$\frac{u_c(y)}{y} = \sqrt{\frac{u_c^2(x_1)}{x_1^2} + \frac{u_c^2(x_2)}{x_2^2}} \qquad (6.31)$$

Equation 6.31 gives a simple rule for combining quantities that are multiplied or divided: the relative uncertainty squared is the sum of the relative uncertainties squared of the components. The values of the quantities are required to calculate the relative uncertainty (the uncertainty divided by the value), and this underlines the fact that measurement uncertainty is a property of a particular result, not a property of a method. Many texts just give equation 6.31 as the multiplication rule, but I thought it might be informative to show that it comes from the underlying equation, as does the equation for added quantities. Equation 6.31 also holds for the uncertainty of $y = x_1 x_2$, but I shall leave the reader to do the more simple mathematics.

### 6.6.4.4  Logarithmic and Exponential Functions

Most analytical measurements conform to simple arithmetic rules that have been covered above. When confronted by other mathematical functions, then equation 6.21 or one of its simplifications can be used. For example, if the change in intensity of a light source (from $I_0$ to $I$) is observed and converted to an absorbance ($A$)

$$A = \log_{10} \frac{I_0}{I}$$

(6.32)

the uncertainty of the absorbance $u(A)$ is given by the following mathematics.

$$dA = \log_{10}(e) \left[ \frac{dI_0}{I_0} + \frac{dI}{I} \right]$$

(6.33)

Application of equation 6.23 gives

$$u^2(A) = \log_{10}(e) \left[ \left( \frac{u(I_0)}{I_0} \right)^2 + \left( \frac{u(I)}{I} \right)^2 \right]$$

(6.34)

Equation 6.34 shows that the absolute uncertainty in absorbance is proportional to the combination of relative uncertainties in the intensities $I_0$ and $I$.

### 6.6.4.5  Spreadsheet Method

If the algebraic manipulations required by equation 6.21 are becoming too complicated, there is a spreadsheet method that gives the answer directly from the uncertainty components and the function for $y$. It relies on the fact that uncertainties are usually only a small fraction of the quantities (a few percent at most), and so the simplifying assumption may be made that, for $y = f(x)$:

$$\frac{dy}{dx} \approx \frac{\Delta y}{\Delta x} \approx \frac{u(y)}{u(x)} = \frac{f(x+u(x))-f(x)}{u(x)} \tag{6.35}$$

and so

$$u(y) = \frac{dy}{dx}u(x) \approx f(x+u(x))-f(x) \tag{6.36}$$

For more than one contributing uncertainty component, the $u(y)$ calculated one $x$ at a time can be squared and summed to give the square of the combined uncertainty.

As an example, consider the measurement of the purity of a chemical by quantitative NMR, using the peak of an added CRM as an internal standard. The purity of the test material ($P_{test}$) is given by

$$P_{test} = \frac{P_{CRM}m_{CRM}I_{test}}{m_{test}I_{CRM}} \tag{6.37}$$

where $m$ are masses and $I$ are peak areas. Suppose the standard uncertainties for each term in equation 6.37 have already been estimated, by algebra or by the spreadsheet method, and they are independent and can be combined by equation 6.23. Spreadsheet 6.3 shows values and standard uncertainties that will be used in the calculation.

In the spreadsheet method (see spreadsheet 6.4) the table is set out with each column having all the parameters for the calculation and one column for each parameter. The table is therefore square with, in the example above, five by five variables ($P_{CRM}$, $m_{CRM}$, $I_{test}$, $m_{test}$, $I_{CRM}$). In each column each of the variables in turn is changed by adding the uncertainty. (These are the shaded cells on the diagonal in spreadsheet 6.4). At the bottom of each column a row is created with the formula (equation 6.37), using the values (including the one value augmented by the uncertainty) in that column. Another column is created at the left of the table with the unchanged values of each variable; the equation cell thus leads to the result of the measurement. At the bottom of each column, the column calculation of the measurand is subtracted from the measurement result in the first (left-hand) column, then the square root of the sum of the squares of these differences is taken (using =SQRT(SUMSQ(range))).

Note that repeatability precision ($r$) is included in the equation for the measurand with a nominal value of 1 and standard uncertainty the Type A standard deviation from repeated measurements.

$$P_{test} = \frac{P_{CRM}m_{CRM}I_{test}}{m_{test}I_{CRM}} \times r \tag{6.38}$$

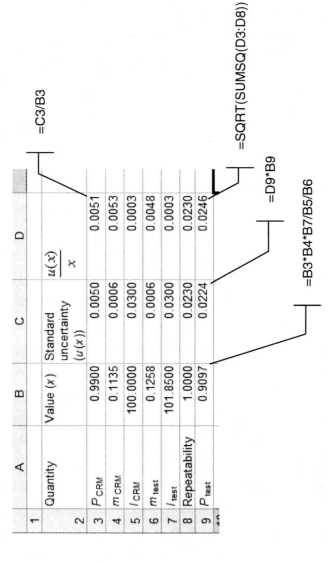

**Spreadsheet 6.3.** Values and uncertainties for quantities used in the calculation of the purity of a sample by quantitative NMR using equation 6.37.

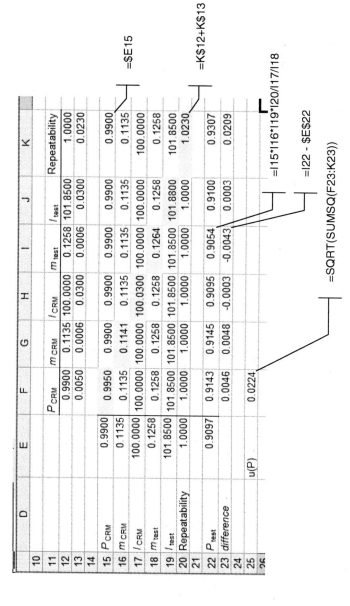

| | $P_{CRM}$ | $m_{CRM}$ | $l_{CRM}$ | $m_{test}$ | $l_{test}$ | Repeatability |
|---|---|---|---|---|---|---|
| | 0.9900 | 0.1135 | 100.0000 | 0.1258 | 101.8500 | 1.0000 |
| | 0.0050 | 0.0006 | 0.0300 | 0.0006 | 0.0300 | 0.0230 |
| | | | | | | |
| $P_{CRM}$ | 0.9900 | 0.9950 | 0.9900 | 0.9900 | 0.9900 | 0.9900 |
| $m_{CRM}$ | 0.1135 | 0.1135 | 0.1141 | 0.1135 | 0.1135 | 0.1135 |
| $l_{CRM}$ | 100.0000 | 100.0000 | 100.0000 | 100.0300 | 100.0000 | 100.0000 |
| $m_{test}$ | 0.1258 | 0.1258 | 0.1258 | 0.1258 | 0.1264 | 0.1258 |
| $l_{test}$ | 101.8500 | 101.8500 | 101.8500 | 101.8500 | 101.8800 | 101.8500 |
| Repeatability | 1.0000 | 1.0000 | 1.0000 | 1.0000 | 1.0000 | 1.0230 |
| | | | | | | |
| $P_{test}$ | 0.9097 | 0.9143 | 0.9145 | 0.9095 | 0.9054 | 0.9100 | 0.9307 |
| difference | | 0.0046 | 0.0048 | -0.0003 | -0.0043 | 0.0003 | 0.0209 |
| | | | | | | |
| u(P) | | 0.0224 | | | | |

=$E15

=K$12+K$13

=I15*I16*I19*I20/I17/I18

=I22 - $E$22

=SQRT(SUMSQ(F23:K23))

**Spreadsheet 6.4.** Spreadsheet method for the measurement uncertainty of the purity of a sample by quantitative NMR.

The result $[u_c(P_{test}) = 0.0224]$ may be compared with that calculated from equation 6.31, which is calculated in spreadsheet 6.3, cell C9, from the relative combined standard uncertainty in cell D9.

$$u_c(\text{test}) = P_{test}\sqrt{\left(\frac{u(P_{CRM})}{P_{CRM}}\right)^2 + \left(\frac{u(m_{CRM})}{m_{CRM}}\right)^2 + \left(\frac{u(I_{test})}{I_{test}}\right)^2 + \left(\frac{u(m_{test})}{m_{test}}\right)^2 + \left(\frac{u(I_{CRM})}{I_{CRM}}\right)^2 + \left(\frac{u(r)}{1}\right)^2}$$

$$= 0.9097 \times \sqrt{0.0051^2 + 0.0053^2 + 0.0003^2 + 0.0048^2 + 0.0003^2 + 0.023^2} \qquad (6.39)$$

$$= 0.0224$$

In fact, there is only a difference in the sixth decimal place.

For straightforward ratio calculations, it is as easy to sum and square relative uncertainty terms as it is to lay out the spreadsheet calculation, but when there are mixed sums and products and nonlinear terms, the spreadsheet method is much easier.

### 6.6.4.6  Correlation

The simple equations used to combine uncertainty rely on the independence of the values of the components. Consider a titration volume calculated from the difference between initial and final readings of a burette and the three uncertainty components identified for volume in table 6.2. Repeatability should be genuinely random, and so the combined uncertainty of a difference measurement with repeatability $u(r)$ is

$$u_c(r) = \sqrt{u(r)^2 + u(r)^2} = \sqrt{2}u(r) \qquad (6.40)$$

Similarly, if the temperature randomly fluctuates during the experiment, the effect can be applied to the initial and final volumes. However, for a constant effect, any calibration error cancels completely, and cancels to some extent for a proportional effect. Thus if a reading of volume $v_{obs}$ is in reality $v_{true} + \Delta v$, the difference between two readings, $v_{obs,1}$ and $v_{obs,2}$, is

$$(v_{obs,2} - v_{obs,1}) = (v_{true,2} + \Delta v) - (v_{true,1} + \Delta v) = (v_{true,2} - v_{true,1}) \quad (6.41)$$

There are often correlations in analytical measurements because ratio or difference calculations are made from data obtained using the same instrument or equipment, with systematic effects that are likely to be unchanged during the two measurements. I develop the theory for a simple ratio with perfect correlation below.

Consider a ratio $R = a/b$, for which the uncertainties of each of $a$ and $b$ are the same and equal to $u$. The uncertainty in $R$ is given by equation 6.21, and if the correlation coefficient $r$ is 1, then

$$u_c^2(R) = \left(\frac{\partial R}{\partial a}\right)^2 u^2 + \left(\frac{\partial R}{\partial b}\right)^2 u^2 + 2\left(\frac{\partial R}{\partial a}\right)\left(\frac{\partial R}{\partial b}\right)u^2 = u^2\left(\left(\frac{\partial R}{\partial a}\right)+\left(\frac{\partial R}{\partial b}\right)\right)^2 \quad (6.42)$$

Working through the algebra gives

$$u_c^2(R) = u^2\left(\frac{a^2}{b^4} + \frac{1}{b^2} - 2\frac{a}{b^3}\right) \quad (6.43)$$

or

$$\frac{u_c^2(R)}{R^2} = \frac{u^2}{a^2}(1-R)^2 \quad (6.44)$$

If complete independence of $a$ and $b$ is assumed, having the same magnitude of uncertainty ($u$), then equation 6.23 holds and

$$\frac{u_c^2(R)}{R^2} = \frac{u^2}{a^2}(1+R^2) \quad (6.45)$$

Equation 6.44 shows that for a ratio of unity ($R = 1$), the correlated uncertainties cancel completely, while the uncorrelated ones give the familiar factor of $\sqrt{2}$.

There is still some debate about what is and what is not correlated. The assumption there is no correlation leads at worst to an overestimate of the uncertainty. Also, because the repeatability, which is always uncorrelated (or should be), is often the greatest component of the combined uncertainty, the contributions of correlated effects are small.

### 6.6.4.7  Repeated Measurements

The uncertainty is a property of a particular measurement result and takes into account whatever information is available. So if a result is the mean of a number of independent determinations, the repeatability contribution to the combined uncertainty is the standard deviation of the mean (i.e., the standard deviation divided by the square root of the number of determinations):

$$u(\bar{x}) = \frac{s_r}{\sqrt{n}} \quad (6.46)$$

If an estimate of the measurement uncertainty has been made and quoted for a single determination, it is possible to infer the measurement uncertainty for repeated determinations by going through the uncertainty budget and dividing appropriate, but not all, terms by $\sqrt{n}$. The systematic effects that have been estimated by the type B procedures usually will not be divided by $\sqrt{n}$. This is an example of the need to properly document an uncertainty budget so that future users of the information can understand what has been done and rework the information if necessary.

### 6.6.4.8  Expanded Uncertainty

The expanded uncertainty is a range about the result in which the value of the measurand is believed to lie with a high level of confidence. It is calculated from the combined standard uncertainty by multiplying by a coverage factor ($k$), and is given the symbol $U$. Although the documented combined standard uncertainty tells everything about the measurement uncertainty, for many purposes, not the least testing against regulatory or contractual limits, the idea of error bars around a result are rather compelling. If the combined standard uncertainty has the properties of a standard deviation of a normally distributed variable, then about 68.3% of the distribution will fall within ± 1 standard deviation of the mean, 95.4% will fall within ± 2 standard deviations and 99.7% will fall within ± 3 standard deviations. So by multiplying the combined standard uncertainty by $k = 2$, we can say that the value of the measurand (equated to the mean of the population) will lie within the range result ± $U$ with 95% confidence.

There is a problem when some components of the combined standard uncertainty are assessed from measurements or estimates with finite degrees of freedom. A type A estimate from a standard deviation of $n$ repeated measurements has $n - 1$ degrees of freedom. Usually Type B estimates will be based on data that have essentially infinite degrees of freedom, but if the standard uncertainty is open to doubt, the effective degrees of freedom can be determined from

$$V_{eff} = 0.5 \times \left(\frac{\Delta u}{u}\right)^{-2} \tag{6.47}$$

where $\Delta u$ is the uncertainty of the uncertainty. Equation 6.47 is graphed in figure 6.8.

Thus, if it is believed that the estimate of an uncertainty component is within 10% of the appropriate value (i.e., $\Delta u/u = 0.1$) then there are 50 degrees of freedom. Degrees of freedom are exhausted when the uncertainty in the estimate reaches about 50%. For many Type B estimates there is no uncertainty in the estimate and $v_{eff}$ is infinite. Having determined the degrees of freedom of each uncertainty component, the effective degrees of freedom for the combined uncertainty is calculated from the Welch-Satterthwaite formula (Satterthwaite 1941), taking the integer value rounded down from

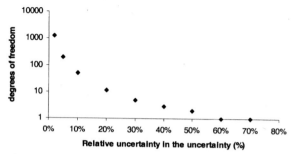

**Figure 6.8.** Effective degrees of freedom from a Type B estimate which has an uncertainty given by equation 6.47.

$$V_{\text{eff}}(u_c) = \frac{u_c^4(y)}{\displaystyle\sum_{i=1}^{i=m} \frac{u_i^4(y)}{V_i}} \tag{6.48}$$

where $u_c(y)$ is the combined standard uncertainty of the result, $y$, and $u_i$ is a component of that uncertainty with degrees of freedom $v_i$. Equation 6.48 weights greater components of the uncertainty and ones with smaller degrees of freedom. Having obtained $v_{\text{eff}}$ for the combined standard uncertainty, the coverage factor $k$ is the point on the two-tailed Student $t$ distribution with a probability $\alpha$ [see table 6.3, or use the Excel formula =TINV($\alpha$, $v_{\text{eff}}$)]. For a 95% confidence interval $\alpha = 0.05$, and in general the percentage of results included in the range is $100 \times (1 - \alpha)$.

When assessing measurement uncertainty as part of a method validation, enough experiments are done to have degrees of freedom that do not adversely affect the coverage factor, and usually $k$ is taken as 2. As long as subsequent field measurements followed the validated method, a measurement uncertainty can be then quoted with $k = 2$. For the most part, therefore, the expanded uncertainty should be calculated from the combined standard uncertainty by

$$U = ku_c \tag{6.49}$$

where $k = 2$.

### 6.6.5   Review Estimates and Report Measurement Uncertainty

The process of review is continual. I have encouraged the rejection (suitably documented and justified) of insignificant components at any stage. In the spreadsheet example of quantitative nuclear magnetic resonance above,

Table 6.3. Values of the two-tailed Student's $t$ distribution for 90%, 95%, and 99% confidence calculated using the Excel function =TINV($\alpha$, degrees of freedom)

| Degrees of freedom | Percentage of the distribution within ± $t\sigma$ | | |
|---|---|---|---|
| | 90% ($\alpha = 0.1$) | 95% ($\alpha = 0.05$) | 99% ($\alpha = 0.01$) |
| 1 | 6.31 | 12.71 | 63.66 |
| 2 | 2.92 | 4.30 | 9.92 |
| 3 | 2.35 | 3.18 | 5.84 |
| 4 | 2.13 | 2.78 | 4.60 |
| 5 | 2.02 | 2.57 | 4.03 |
| 6 | 1.94 | 2.45 | 3.71 |
| 7 | 1.89 | 2.36 | 3.50 |
| 8 | 1.86 | 2.31 | 3.36 |
| 9 | 1.83 | 2.26 | 3.25 |
| 10 | 1.81 | 2.23 | 3.17 |
| 20 | 1.72 | 2.09 | 2.85 |
| 50 | 1.68 | 2.01 | 2.68 |
| 100 | 1.66 | 1.98 | 2.63 |
| Infinity | 1.64 | 1.96 | 2.58 |

the integrated peak areas with uncertainty of the order of 0.3% clearly could have been omitted earlier. But now we are at the end of our endeavors. What is in, is in, and what has been left out is out. A combined standard uncertainty of a measurement result will be communicated to the client. Does it make sense? A small uncertainty might be a source of pride in a careful analysis, but if it is clearly less than the range of results found in practice, something has not been considered or has been underestimated. The repeatability of the measurement is the main intralaboratory component, and the final combined uncertainty has to be greater than the repeatability. For simple operations, there may not be a great deal to add, but for analyses of a usual level of complexity, involving perhaps sampling, sample preparation, recovery corrections, and calibration with a CRM, the combined uncertainty will be two to three times the within laboratory repeatability.

As reproducibility standard deviation from interlaboratory method validation studies has been suggested as a basis for the estimation of measurement uncertainty if it is known $s_R$ can be compared with a GUM estimate. It may be that with good bias correction, the estimate may be less than the reproducibility, which tends to average out all systematic effects including ones not relevant to the present measurement. Another touchstone is the Horwitz relation discussed in section 6.5.4. A rule of thumb is that the reproducibility of a method (and therefore the estimated measurement uncertainty) should fall well within a factor of two of the Horwitz value.

### 6.6.5.1 Displaying Uncertainty Components

It might be useful to have a picture of the contributions of effects, and this can be done by a Pareto or other form of bar graph. Although a client is unlikely to need these graphs, they are a useful tool when assessing the uncertainty budget and can be used in summaries for quality managers and accreditation bodies.

Consider the example of quantitative NMR. Spreadsheet 6.3 gives the standard uncertainties and relative standard uncertainties of the components of the combined uncertainty. It is usual to graph the relative standard uncertainties, the standard uncertainties multiplied by the sensitivity coefficient $[\partial y/\partial x \; u_c(x)]$, or the squares of the latter expressed as a percentage contribution to the combined uncertainty (see equation 6.23). A horizontal bar chart for each component in decreasing order is one way of displaying these values (figure 6.9).

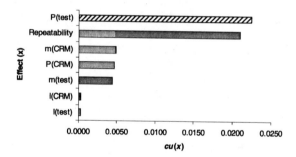

(a)

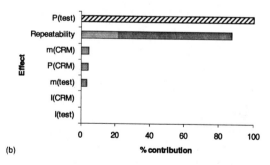

(b)

**Figure 6.9.** Bar charts of the uncertainty components in the quantitative NMR example.

(a) The value of $c \; u(x) = \left| \dfrac{\partial P_{test}}{\partial x} u(x) \right|$ for each component with the total uncertainty $P_{test}$ (hatched bar). (b) The percent contribution of each uncertainty component.

A Pareto chart is used to display effects that sum to 100%. Bars represent individual effects, and a line is the cumulative effect (figure 6.10). The effects are ordered from the greatest to the least, and often show the Pareto principle that 20% of the effects contribute 80% of the uncertainty.

### 6.6.5.2 Reporting Measurement Uncertainty

The final combined standard uncertainty, whether obtained using algebra or a spreadsheet or other software, is the answer, and can be quoted as such. I recommend using the wording: "Result: x *units* [with a] standard uncertainty [of] $u_c$ *units* [where standard uncertainty is as defined in the International Vocabulary of Basic and General Terms in Metrology, 3rd edition, 2007, ISO, Geneva, and corresponds to one standard deviation]."

The words in brackets can be omitted or abbreviated as appropriate. I do not recommend that you use plus or minus (±)with a combined uncertainty. If an expanded uncertainty is to be quoted, then ± can be used because it does define a probability range. The result should be recorded as: "Result: (x ± U) *units* [where] the reported uncertainty is [an expanded uncertainty as defined in the International Vocabulary of Basic and General Terms in Metrology, 3rd edition, 2007, ISO, Geneva] calculated with a coverage factor of 2 [which gives a level of confidence of approximately 95%]." Although explanations of expanded uncertainty and the meaning of the probability level may be omitted, the coverage factor must be included. It should not be automatically assumed that $k = 2$.

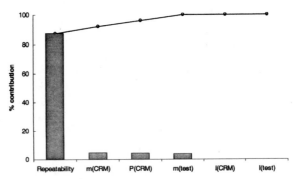

**Figure 6.10.** Pareto chart of the contributions to the uncertainty in the quantitative NMR example. Bars: individual contributions; line: cumulative uncertainty.

## 6.7 Conclusions

I close this chapter with a discussion of what measurement uncertainty means and how it relates to repeatability and other precision measures. What is the difference between $30 \pm 1$ nmol $L^{-1}$ where 30 nmol $L^{-1}$ is the mean and $\pm 1$ µg $g^{-1}$ is the 95% confidence interval calculated from the standard deviation ($s$) of five repeats of the experiment ($= t_{0.05",4} \times s/\sqrt{5}$), and $30 \pm 1$ nmol $L^{-1}$ where 30 nmol $L^{-1}$ is a single result and $\pm 1$ nmol $L^{-1}$ is the expanded uncertainty with infinite degrees of freedom from a full uncertainty budget? To help with the example, assume that the measurement is of cadmium in water, for which the World Health Organization (WHO) upper limit is 27 nmol $L^{-1}$. In the first case a correct statement would be "The level of cadmium in the test sample of water was $30 \pm 1$ nmol $L^{-1}$ (95% confidence interval, $n = 5$). Given the result of the chemical analysis and its associated repeatability, the odds that other measurements of the test sample, conducted under identical conditions, would determine that the concentration of cadmium did not exceed the WHO guideline of 27 nmol $L^{-1}$ is 1:880 against." Note that there is nothing that can be said about the value of the concentration of cadmium in the sample, just what might happen if the sample were reanalyzed. In contrast, if the figure quoted is the measurement uncertainty, correctly estimated, the statement might be, "The level of cadmium in the test sample of water discharged by the defendant's factory was $30 \pm 1$ nmol $L^{-1}$ (1 nmol $L^{-1}$ is the expanded uncertainty of the result with coverage factor 2, which gives an interval covering approximately 95% of values attributable to the measurand). Given the result of the chemical analysis and its associated uncertainty, the odds that the concentration of cadmium in the sample tested was below the WHO guideline of 27 nmol $L^{-1}$ was 1: 500 million against." The great difference in the odds comes from the degrees of freedom of the uncertainty.

## References

EURACHEM (2000), Quantifying uncertainty in analytical measurement, 2nd ed., M. Rosslein S.L.R. Ellison, A. Williams (eds.) (EURACHEM/CITAC, Teddington UK).

Hibbert, D B and Gooding, J J (2005), *Data analysis for chemistry: an introductory guide for students and laboratory scientists* (New York: Oxford University Press).

Horwitz, W (1995), Revised protocol for the design, conduct and interpretation of method performance studies. *Pure and Applied Chemistry*, 67, 331–43.

Horwitz, W (2003), The certainty of uncertainty. *Journal of AOAC International*, 86, 109–11.

Horwitz, W, Jackson, T, and Chirtel, S J (2001), Examination of proficiency and control recovery data from analyses for pesticide residues in food: sources of variability, *Journal of AOAC International*, 84 (3), 919–35.

Horwitz, W, Kamps, L R, and Boyer, K W (1980), Quality assurance in the analysis of foods for trace constituents. *Journal of AOAC International*, 63 (4), 1344–54.

ISO (1993a), *International vocabulary of basic and general terms in metrology*, 2nd ed. (Geneva: International Organization for Standardization).

ISO (1993b), Guide to the expression of uncertainty in measurement, 1st ed. (Geneva: International Organization for Standardization).

ISO (1994), Precision of test methods—Part 1: General principles and definitions, 5725-1 (Geneva: International Organization for Standardization).

ISO/IEC (2005), General requirements for the competence of calibration and testing laboratories, 17025 2nd ed. (Geneva: International Organization for Standardization).

Joint Committee for Guides in Metrology (2007), *International vocabulary of basic and general terms in metrology*, 3rd ed. (Geneva: International Organization for Standardization).

Satterthwaite, F E (1941), Synthesis of variance. *Psychometrika*, 6 (5), 309–17.

Thompson, M (2005), AMC technical brief 17: The amazing Horwitz function. Available: http://www.rsc.org/pdf/amc/brief17.pdf.

# 7

## Metrological Traceability

### 7.1 Introduction

The ability to trace a measurement result to a reference value lies at the heart of any measurement. Traceability is part of standards governing laboratory practice, such as ISO/IEC 17025 and Good Laboratory Practice (see chapter 9), as a mandatory property of a measurement result, yet as a concept, traceability of a chemical measurement result is poorly understood. It is either taken for granted, often without much foundation, or ignored altogether. Why is traceability so important? How have we been able to ignore it for so long? The International Union of Pure and Applied Chemistry (IUPAC) has applied itself to this problem and a definitive discussion on metrological traceability in chemistry will be published.[1]

#### 7.1.1 A Note about Terminology

In this chapter I use the term "metrological traceability" to refer to the property of a measurement result that relates the result to a metrological reference. The word "metrological" is used to distinguish the concept from other kinds of traceability, such as the paper trail of documentation, or the physical trail of the chain of custody of a forensic sample. When the term "traceable standard" is used to refer to a calibration material, for example, the provenance of the material is not at issue, but the quantity value embodied in the standard.

### 7.1.2 The Importance of Metrological Traceability

In explaining the importance of metrological traceability, I return to the discussions about chemical measurement (chapter 1). The concentration of a chemical is never measured for its own sake, but for a purpose, which often involves trade, health, environmental, or legal matters. The ultimate goal is achieved by comparing the measurement result with another measurement result, with a prescribed value, a legal or regulatory limit, or with values amassed from the experience of the analyst or client (figure 7.1).

In trading grain, for example, if exported wheat is analyzed by both buyer and seller for protein content, they should be confident that they will obtain comparable measurement results; in other words, results for the same sample of wheat should agree within the stated measurement uncertainties. If the results do not agree, then one party or the other will be disadvantaged, the samples will have to be remeasured, perhaps by a third-party referee, at cost of time and money. Different results could arise from genuine differences in the value of the quantity; for example, the wheat could have taken up moisture or been infected with fungus, so it is important that the analysis does not lead to incorrect inferences.

The need for comparability also extends in time. To understand temporal changes of a system being monitored, such as the global atmosphere, results obtained at one time must be comparable with those obtained at

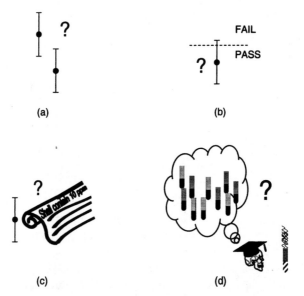

**Figure 7.1.** Comparisons among measurement results: (a) with other results, (b) with limits, (c) with an expected value, and (d) with experience.

another time, in the same or in another laboratory. Carbon dioxide levels in the atmosphere have been measured since the 1960s at Mauna Loa, Hawaii, and it is these results that provide evidence of global warming (Emanuele et al. 2002). Before political decisions are made, there must be confidence that the difference between results in 1970 and in 2000 arises from genuine changes in atmospheric carbon dioxide and not the inevitable changes in instruments, personnel, and even methods of analysis. This is assured, even when calibrators or measurement systems are different, when the results are traceable to the same stated metrological reference maintained through time.

### 7.1.3   Metrological Traceability Defined

The *International Vocabulary of Basic and General Terms in Metrology* (VIM), second edition (ISO 1993, term 6.10), defines traceability as the "property of the result of a measurement or the value of a standard whereby it can be related to stated references, usually national or international standards, through an unbroken chain of comparisons all having stated uncertainties." The third edition (Joint Committee for Guides in Metrology 2007) uses "calibrations" instead of "comparisons" and makes the point that these contribute to the measurement uncertainty. The third edition of VIM also distinguishes metrological traceability from other kinds of traceability.

- Metrological traceability is a property of a measurement result, not a method or measurement procedure, but the result itself.
- A metrological reference is a document defining (usually) a unit. A rational quantity scale, of quantities that can be compared by ratio and have a zero (e.g., length, mass, amount of substance, but not temperature in degrees Celsius), must have a unit, even if it is 1.
- An unbroken chain of comparisons/calibrations means that there must be an identified path of assignments of value to calibrators from the calibrator used in the final measurement back to the *primary calibrator* that is the embodiment (also known as the realization) of the unit.
- Measurement uncertainty of the value of the calibrators increases down the chain, and although, if properly chosen, the measurement uncertainty of the value of the calibrator might be small compared with other uncertainties of the measurement, the measurement uncertainty of the value of the calibrator is an essential ingredient of the metrological traceability of the result.

Perhaps this sounds unnecessarily complicated, but an understanding of basic concepts and terms in metrology help us appreciate the importance of metrological traceability, measurement uncertainty, and the like.

A direct consequence of metrological traceability is that if two measurement results are metrologically traceable to the same metrological reference, then they must be comparable (i.e., can be compared) within their measurement uncertainties. This is sometimes explained in terms of the futility of trying to compare apples and oranges, or in its positive sense, comparing "like with like."

## 7.2 Comparability of Measurement Results

The foregoing discussion identifies the need for comparability and its achievement by metrological traceability. Comparability of measurement results is conveniently defined in terms of metrological traceability. If two results are traceable to the same stated metrological reference, then they must be comparable. Please note that in common speech "comparable" often means "about the same magnitude," but this is not the case here. In the laboratory "comparability of measurement results" means simply that the results can be compared. The outcome of the comparison, whether the results are considered near enough to be equivalent, is not a factor in comparability here.

To show how comparability and traceability work in practice, take the example of a mass measurement. If I weigh myself on a bathroom scale, the result is around 77 kg. These scales have been calibrated in the factory as part of the manufacturing process using standard masses, which are periodically calibrated against an Australian standard mass provided by a calibrating authority. This mass in turn is calibrated against a standard mass held at the National Measurement Institute in Sydney. Metrological traceability to the international standard is assured when this mass is compared with a secondary standard kilogram at the International Bureau of Weights and Measures (BIPM) in Sèvres near Paris. Occasionally this secondary standard is, with great ceremony and following procedures outlined in an international treaty, compared to the international prototype of the kilogram. This artifact is a piece of platinum iridium alloy that serves as the mass standard in the international system of units (SI). Under defined conditions of preparation and use, this lump of metal has a mass of 1 kg with no uncertainty and the authority of international treaty. The bathroom scale is not very accurate. I can obtain different readings on different floor surfaces and at different times. My guess is that the 95% confidence interval on my mass is at least ± 2 kg. At my doctor's office the scales give my mass as 76.3 kg, with an uncertainty of 0.3 kg. These scales have been calibrated with masses that can also be traced to the mass of the SI kilogram, but not necessarily by the same route. Nonetheless, being traceable to the same stated reference renders the bathroom scales and the doctor's scales, within measurement uncertainty, comparable.

As both measurements are made in Australia and the weighing instruments (bathroom and doctor's scales) are calibrated to the mass of the Australian standard kilogram, the traceability chain could have stopped there, along with any comparability. Although the definitions do not mention the SI, and of course for some measurements there are no SI units anyway, the more universal the stated reference is, the more measurements will come under its aegis. Below I discuss some of the scenarios for comparison shown in figure 7.1 in more detail, stressing the importance of measurement uncertainty. I will also explore the nature of the arrows that define the metrological traceability in figure 7.2.

### 7.2.1 Comparing One Result against Another

Measurement results that are traceable to the same reference (often but not necessarily an SI unit), can only be compared in relation to their uncertainties. Figure 7.3 shows the comparison of two measurement results. Because it is unlikely that two measurements made of the same material will give identical results, it is only by evaluating the measurement uncertainties that the client can decide whether he or she will accept that the two materials are equivalent, in terms of the quantity measured.

The first criterion is that the measurements must be made in the same units—that is, traceable to the same metrological reference. This sounds obvious, but without attention to metrological traceability of a measurement

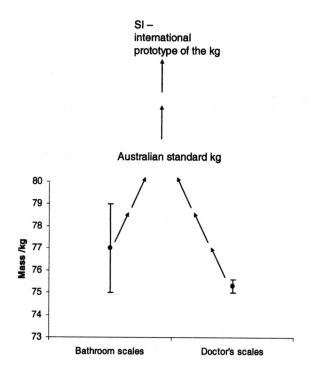

**Figure 7.2.** Traceability of mass measurements to the Australian standard kilogram and to the international prototype of the kilogram.

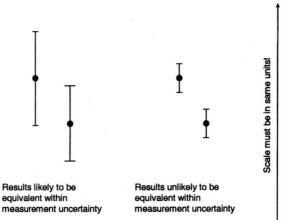

Results likely to be
equivalent within
measurement uncertainty

Results unlikely to be
equivalent within
measurement uncertainty

**Figure 7.3.** Comparison of test results given a
result and 95% confidence interval. (a) There is a
high probability that the value of each measurand
is the same. (b) There is only a low probability
that the value of each measurand is the same.

result, it is possible to quote the same units (mol L$^{-1}$, for example) but actually not to obtain a measurement result that is metrologically traceable to the
same definition of the unit in question. Having made the measurements and
estimated the measurement uncertainties the results ($x_1$ and $x_2$) can be compared by computing the probability of the hypothesis $H_0$: $\mu_1 = \mu_2$ (i.e., equality
of the [true] values of the measurand in each experiment). If the measurement
uncertainty of each of two independent measurements is expressed as a combined standard uncertainty ($u_c = u_1, u_2$), then $H_0$ is tested by computing the
probability that the test statistic $T$ is exceeded $\Pr(T>t)$, where

$$t = \frac{|x_1 - x_2|}{\sqrt{u_1^2 + u_2^2}} \tag{7.1}$$

at an appropriate number of degrees of freedom (see chapter 6). In the
case of infinite degrees of freedom, the Student's $t$ statistic is replaced by
the normal distribution $z$ score, calculated by the same equation. In Excel,
the probability associated with the $t$ value of equation (7.1) is given by
=TDIST($t$, $f$, 2) where $f$ is the degrees of freedom and the 2 indicates
a two-tailed probability (i.e., there is no particular reason to assume $\mu_1$ is
greater or lesser than $\mu_2$). For infinite degrees of freedom Excel requires
=NORMSDIST($-t$)*2. The negative value of $t$ ensures the cumulative distribution is taken from $-\infty$ to $-t$ (i.e., the left-hand tail of the distribution, and
multiplying by two gives the two-tailed probability). If the measurement

uncertainty is given as an expanded uncertainty giving a coverage with probability 95%, $U$, then $U$ is divided by the coverage factor (usually 2) to give $u_c$, and the procedure is continued as above with infinite degrees of freedom. As with all probabilities, it is up to the user to decide which probabilities warrant concern, and this in turn is assessed by considering the risks of making an error, either by accepting $H_0$ when it is actually false, or rejecting $H_0$ when it is true.

### 7.2.2  Comparing a Result against a Limit

Similar arguments apply when considering a result against a limit. The limit may be a legal requirement that must be achieved or not exceeded, or it may be an internal specification. The risks of noncompliance might require evidence that the probability the test item does not comply is very small and insignificant ("positive compliance," line a in figure 7.4) or, alternatively,

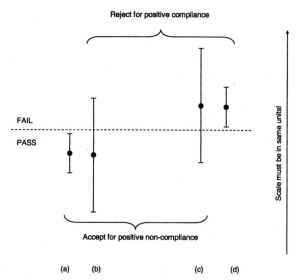

**Figure 7.4.** Comparison of a test result, including uncertainty, against a limit. (a) The result indicates that the material tested complies with a low probability that it does not comply. (b) The result indicates compliance, but the uncertainty leads to a significant probability that the tested material does not comply. (c) The result indicates noncompliance, but the uncertainty leads to a significant probability that the tested material does, in fact, comply. (d) The result indicates that the material tested does not comply with a low probability that, in reality, it does comply.

unless the item clearly does not comply ("positive noncompliance," line d in figure 7.4), then action will not result. In the former case, line a is acceptable and lines b, c, and d are not. In the latter case of positive noncompliance lines a, b, and c are all acceptable, and only line d would lead to action as a result of the analysis. Much of the time, no specific direction is given, so although line a would pass and line d would fail, the two cases in the middle lead to quandary. The example in section 6.7 shows that the nature of the uncertainty quoted determines the statement that can be made, whether about the value of the measurand or about the likely results of repeated experiments.

## 7.3 How Metrological Traceability Is Established

From the definitions and examples above, it should be clear that metrological traceability is established by a series of comparisons back to a reference value of a quantity. In the mass example it is easy to picture a number of scientists with ever more sophisticated balances comparing one mass with another down the chain of masses from the international prototype of the kilogram to bathroom scales. This is known as a calibration hierarchy. A mass farther up the chain is used to calibrate the next mass down the chain, which in turn can be used to calibrate another mass, and so on, until the final measurement of the mass is made.

Demonstrating traceability back to the SI requires identification of the traceability chain. A formal description of the chain might look like figure 7.5. In this figure, the middle set of arrows represents the alternate actions of calibration and assignment. They connect square boxes containing measuring systems on the right and test items on the left. The measuring systems (instruments, balances, glassware) are informed by measurement procedures on the extreme right. The actions result in a measurement result (i.e., the value of the measurand), including appropriate units and uncertainty. The down arrow shows the extent of the calibration hierarchy, and the up arrow between measurement result and metrological reference is the metrological traceability chain. This is a metrologist's view of the matter, and practical approaches will be discussed later.

At each level in the hierarchy, the uncertainty about the quantity value must be stated. What about a chemical measurement, such as the amount of protein in a grain sample or the pH of a liquid sample? The principle is the same. The measuring instrument (e.g., Dumas nitrogen analyzer, glass pH electrode) must be calibrated with a traceable reference material. In many cases in chemistry this is a certified reference material (CRM) or standard calibrated from a CRM. By definition, a CRM has one or more quantity values that have been established with stated uncertainty and metrological traceability. Using a CRM to calibrate a working standard or using a CRM directly for a measurement allows one to claim that the final result is met-

measurement function for end-user's quantity value of the measurand :
$$m_{sample} = f(m_{calibrator\ 6}, p, \theta)$$

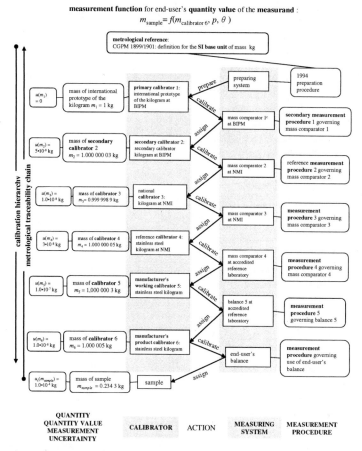

**Figure 7.5.** Metrological traceability of mass measured by bathroom scales.

rologically traceable. One reason that CRMs are expensive is that they must unequivocally demonstrate metrological traceability. This can be time consuming and costly.

### 7.3.1 The Top of the Chain: The Metrological Reference

At the top of the traceability chain is the stated metrological reference, which for our purposes is the definition of a unit. It might not be obvious how a piece of paper from Paris giving the interesting, but rather esoteric, definitions of the SI base units can be used as a reference. The metrological reference allows the creation of a *primary calibrator* that embodies the value of the unit (or some multiple or submultiple of it). The international prototype of the kilogram is an example of such an embodiment, as is a Josephson junction that embodies the volt. These are artifacts that can be used as calibrators to pass on the embodiment to another artifact (secondary calibrator), and so on down the chain.

Confusion about metrological traceability results from the idea that it is possible to trace a measurement result to an object per se, or even to an institution. You may read that a measurement was "traceable to NIST" (the National Institute of Standards and Technology). Leaving aside that the measurement *result* might or might not be traceable, presumably this statement means that a calibration has been performed using a material that has been standardized and sold by NIST, the certificate of which will detail the nature of the metrological traceability. The important bit is the certificate describing traceability. All that comes from NIST is not traceable, and all that is traceable is not solely from NIST.

The international prototype of the kilogram is a simple example of the top of the chain. The object that has a mass of exactly 1 kg when used in accordance with the 1994 definition of the measurement procedure is what everyone thinks of as a standard. It may be accepted that the weights and measures inspector who puts a standard 1 kg on the scales in the local supermarket is using an object whose mass value has been established by a series of comparisons of the kind shown in figure 7.5, but this is not a typical chain because the SI unit of mass is the only remaining base unit established by reference to an artifact (i.e., the piece of Pt/Ir alloy in Sèvres). All the other units are realized by a reference procedure. For example, the unit of time, the second, is defined as the "duration of 9,192,631,770 periods of the radiation corresponding to the transition between the two hyperfine levels of the ground state of the cesium 133 atom," and the unit of length, the meter, is "the distance traveled by light in a vacuum during a time interval of 1/299,792,458 of a second." (BIPM 2006) Why has the length of a bar of platinum, been superseded by this complex definition? The problem with an artifact standard is that there can only be one, and so the world must beat a path to its door, so to speak, in order to compare their national standards with the global standard. This is precisely what happens at present with the kilogram. In contrast, anybody with an atomic clock can realize the second, and therefore the meter. As long as the proper procedure is followed to create embodiments of these units, metrological traceability can be guaranteed without any physical interaction with an international standard object. This is such a powerful argument that the days of the prototype kilogram are decidedly numbered. Remember the definition of the mole you learned in high school? The mole is "the amount of substance of a system which contains as many elementary entities as there are atoms in 0.012 kg of carbon-12." There is an international effort, known as the Avogadro project, to measure the Avogadro constant, involving very precise measurements on a number of silicon spheres in laboratories around the world (Becker 2003; NPL 2005). When the Avogadro constant is known to a suitable degree of precision (about 2 parts in $10^8$), the kilogram can be redefined as the mass of $(6.022 \ldots \times 10^{23}/0.012)$ atoms of carbon-12, and the prototype kilogram will be consigned to a museum. (Physicists have hit back with a definition based on the Planck constant, but that is another story.)

The top of the metrological traceability chain is given in figure 7.6. The procedure and system referred to in the figure may be either for the production of the calibrator, such as the international prototype of the kilogram, or for a primary measurement procedure governing a measuring system, as in the use of a coulometric titration.

### 7.3.2 Primary Measurement Standards and Primary Measurement Methods

The top of the metrological traceability chain resides with artifacts or procedures that realize a unit. These are described in more detail below.

#### 7.3.2.1 Primary Measurement Standards and Calibrators

The international prototype kilogram is an example of a primary measurement standard: its value and uncertainty have been established without relation to another standard of the same kind. When the primary measurement standard is used for calibration, it becomes a primary calibrator.

Primary standards in chemistry are often pure materials that are made and then certified as to identity and purity, sometimes by measuring all possible impurities (organic, inorganic and often specifically water or other solvents) and subtracting from 100%. The unit of the measurement of purity is usually kg/kg = 1 (i.e., the mass of the designated compound or element per kilogram of CRM).

An example of a primary procedure to establish the mass fraction of a pure reference material is the use of coulometry to prepare an amount of an

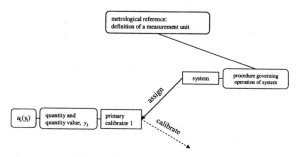

**Figure 7.6.** The top of a metrological traceability chain.

element. If a current $I$ A is passed for $t$ seconds through a solution of silver nitrate, and the only reaction at the cathode is the electrodeposition of silver by

$$Ag^+ (aq) + e \rightarrow Ag (s) \tag{7.2}$$

then the amount of silver deposited is $I(t)/F$, where $F$ is the Faraday constant (96485.3383 C mol$^{-1}$ with $u = 0.0083$ C mol$^{-1}$). If the current source and measurement and the time measurement are traceable, then the amount of silver is traceable.

Primary calibrators, indeed all calibrators, must be commutable; that is, they must behave during measurement in an identical manner to the native analyte material being measured. Matrix reference materials made by mixing a pure reference material with the components of the matrix are unlikely to be entirely commutable, and for this reason some authorities (EURACHEM for one [EURACHEM and CITAC 2002]) advise against using matrix-matched CRMs for calibration, recommending instead their use to establish recoveries, after calibration by a pure reference standard.

A primary measurement standard is expected to have a known quantity value with minimum uncertainty, although this is not specified as such in the definitions. Because the measurement uncertainty is propagated down the calibration hierarchy, you should start with as small an uncertainty as possible and choose methods to establish purity of primary reference materials with this in mind.

### 7.3.2.2  Primary Measurement Procedures

Primary measurement procedures can be at the top of the calibration hierarchy, as they entail how the definition of the unit is realized. I have discussed primary measurement procedures in relation to time and length. The purpose of a primary measurement procedure is to obtain the quantity value and measurement uncertainty of a primary measurement standard. For non–SI units, and sometimes for systems of measurements for which consistency among a group of laboratories is important, adopting a common measurement procedure is a way to provide comparability. The wider the group, the greater the extent of comparability. The procedure needs to be well validated and proven to be sufficiently robust in diverse laboratories. This is usually accomplished by periodic interlaboratory trials (chapter 5). Such a procedure is known as a reference measurement procedure, and it can furnish the stated metrological reference in the definition of metrological traceability.

### 7.3.3  Multiple Chains

The metrological traceability of a measurement result is rarely as straightforward as the chain of comparisons shown in the example of mass in figure 7.2. This example hides the control of influence factors such as temperature and the effect of buoyancy (which requires a knowledge of the density of the mass being weighed). A chemical measurement involving amount-of-substance concentration invariably leads to mass and volume measurements and the attendant effect of temperature. As soon as a measurement is made that gives an input quantity to the measurement result (for an example, see the preparation of a calibration solution of cadmium ions in section 7.3.4.1), the result of this measurement must be traceable. In the case of independent mass and volume measurements leading to a mass concentration, the two traceability chains could be diagrammed with equal importance to their respective metrological references. Where the input quantity is either an influence quantity (i.e., one not directly in the measurement function) or a conversion factor such as molar mass, its chains might be drawn as side chains. There is no difference in the requirement for metrological traceability, but it is likely that the traceability of these quantities will have been established earlier, and a brief statement to this effect is all that is required when documenting the metrological traceability. For example, reference to the laboratory standard operating procedures that specify the calibration regime of balances, the use of calibrated glassware and thermometers, and the use of the latest IUPAC atomic weights and published fundamental constants, with their measurement uncertainties, is sufficient. An attempt to convey the multiplication of chains in the example of a titration is shown in figure 7.7.

### 7.3.4   Establishing Metrological Traceability of Amount-of-Substance Measurements

The definition of the mole requires statement of the system (see section 7.3.1). As a result there is the potential for a nearly infinite number of amount standards. Regarding length, producing something that is 1 m suffices as a realization of the unit of length. In contrast, a piece of pure silver with mass of 107.8682 g is the embodiment of a mole of silver atoms, but it is not a mole of anything else. Chemists usually interact with amount-of-substance via mass. The mass of a compound of known (or assumed) identity and purity is used to calculate the amount of substance by dividing by the molar mass. Under the auspices of the International Union of Pure and Applied Chemistry, atomic weights of elements and therefore molar masses of compounds have been measured with metrological traceability to the SI kilogram and mole. A pure reference material comes with a certificate of metrological traceability of its identity and purity. When a certain mass of the standard is dissolved in a volume of solvent, the resulting concentration is traceable if the mass measurement and the volume of the flask were themselves traceable. The volume of standardized glassware is only certified at a particular temperature, and so temperature measurements must be made to ensure the

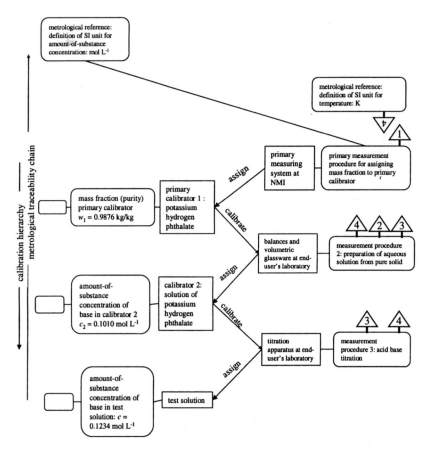

**Figure 7.7.** Multistranded traceability chains, showing the need to establish traceability to a number of references in a chemical measurement (here titration).

defined (traceable) temperature, to make corrections based on traceable temperature measurements, or to expand the uncertainty to allow for possible temperature fluctuations over a given temperature range. Below I give an example of how a calibration solution is prepared.

To prepare a calibration solution of 1 $\mu$mol L$^{-1}$ of Cd$^{+2}$ from a CRM of pure Cd metal, a mass of Cd is weighed into a flask, dissolved, and made up to the mark. The atomic weight of cadmium is 112.411 g mol$^{-1}$ with $u_c = 0.008$ g mol$^{-1}$. Thus, 1.12411 g of pure cadmium metal dissolved in hydrochloric acid and made up to 1.00 L will have an amount concentration of 1.00 × 10$^{-2}$ mol L$^{-1}$. An aliquot of 0.100 mL of this solution made up to 1.00 L should create the desired solution of amount concentration 1.00 $\mu$mol L$^{-1}$. Until a proper estimate of the measurement uncertainty is made the significant figures quoted are not substantiated. The EURACHEM guide to measurement uncertainty (EURACHEM 2000) addresses the uncertainty of the preparation of a 1 g L$^{-1}$ cadmium solution. The components of uncertainty are the purity of the metal used and the weighings and volumes of glassware. The combined standard uncertainty is about 0.9 mg L$^{-1}$. In the preparation of a standard solution, the uncertainty in the atomic weight of cadmium and the extra dilution step (which includes two volumes, the volume pipetted and the volume of the second 1-L volumetric flask) must also be considered. The measurand, the amount concentration ($c_{Cd}$), is related to the input quantities by

$$c_{Cd} = \frac{P \times m}{M \times V_1} \times \frac{V_2}{V_3}$$
(7.3)

where $P$ is the purity, $m$ is the mass of cadmium, $M$ is the atomic weight of cadmium, $V_1$ and $V_3$ are the 1-L dilution volumes, and $V_2$ is the 0.1-mL aliquot of the first solution that is further diluted. Using the methods of chapter 6, the relative standard uncertainty of the calibration solution is calculated as 0.0013, which leads to the amount concentration and its 95% confidence interval, 1.0000 ± 0.0026 µmol L$^{-1}$ (or 1.000 ± 0.003 µmol L$^{-1}$). In using equation 7.3 it is assumed that the temperature of the laboratory has been monitored sufficiently during the procedure to use the calibrated volumes of the glassware in equation 7.3; that is, the uncertainty in the volumes due to uncertainty in the temperature properly reflects the range of temperatures experienced during the preparation.

Is the concentration of this solution metrologically traceable? To be traceable, each of the input quantities in equation 7.3 must be metrologically traceable. These input quantities are listed in table 7.1.

Temperature is only used here in estimating uncertainty, so it does not have the same impact as direct input quantities. As long as the estimate of the range of temperatures used in calculating uncertainty is reasonable, the exact traceability of the temperature should not be a major concern. Even if correction factors were introduced into equation 7.3,

$$c_{Cd} = \frac{P \times m}{M \times V_1 \times f(T_1)} \times \frac{V_2 \times f(T_2)}{V_3 \times f(T_3)}$$
(7.4)

$$f(T) = 1 - 0.00021 \times (T / °C - 20)$$
(7.5)

when temperature is now an explicit input quantity, imprecise knowledge of the temperature still does not lead to great uncertainty in the value of the measurand, and the scrutiny of the metrological traceability of the temperature might not be as great as it is for the purity of the cadmium metal, for instance.

Table 7.1. Metrological traceability of the input quantities to a cadmium calibration solution containing $1.000 \pm 0.003$ $\mu$mol L$^{-1}$ Cd$^{2+}$

| Input quantity | Units (metrological reference) | Relative standard uncertainty | Traceability statement |
|---|---|---|---|
| Purity of Cd metal, $P$ | g/g (SI unit)[a] | $6 \times 10^{-5}$ | CRM certificate with Cd metal; material used as described |
| Atomic weight of Cd, $AW$ | g mol$^{-1}$ (SI units)[a] | $7 \times 10^{-5}$ | Statement by IUPAC describing measurement of the atomic weight and assignment of uncertainty |
| Mass of Cd, $m$ | g (SI unit)[a] | $5 \times 10^{-4}$ | Certificate of calibration of balance giving period of validity of calibration and uncertainty |
| Volumes, $V_1$, $V_3$, $V_3$, | L (SI unit)[a] | $7 \times 10^{-4}$ | Manufacturers certificate of volume of glassware giving temperature at which the volume is valid and assignment of uncertainty |
| Temperature | °C (SI unit)[a] | Used in the standard uncertainty of $V$ | Certificate of calibration of thermometer used to monitor laboratory temperature |

[a] g, L, and °C are derived units in the SI, mol is a base unit.

### 7.3.5 Propagation of Measurement Uncertainty Through a Metrological Traceability Chain

For a measurement result to be metrologically traceable, the measurement uncertainty at each level of the calibration hierarchy must be known. Therefore, a calibration standard must have a known uncertainty concerning the quantity value. For a CRM this is included in the certificate. The uncertainty is usually in the form of a confidence interval (expanded uncertainty; see chapter 6), which is a range about the certified value that contains the value of the measurand with a particular degree of confidence (usually 95%). There should be sufficient information to convert this confidence interval to a standard uncertainty. Usually the coverage factor ($k$; see chapter 6) is 2, corresponding to infinite degrees of freedom in the calculation of measurement uncertainty, and so the confidence interval can be divided by 2 to obtain $u_c$, the combined standard uncertainty. Suppose this CRM is used to calibrate a method, and the uncertainty budget for a measurement, not including the contribution of the CRM, gives a value of $u_{meas}$ for a single measurement, then the combined uncertainty of the result is

$$u_c(\text{result}) = \sqrt{u_{meas}^2 + u_{CRM}^2} \qquad (7.6)$$

A good CRM will have $u_{CRM} \ll u_{meas}$, so its use should not add greatly to the measurement uncertainty. If a CRM is used to create working standards in house, the cost savings is balanced against increasing the measurement uncertainty of the final measurements. If the measurement result for which the measurement uncertainty is given by the $u_c$ in equation 7.6 were the standardization of the working in-house reference material (call it $u_{workingRM}$), then its use in an analysis increases the uncertainty of that measurement by about a factor of $\sqrt{2}$:

$$u_c = \sqrt{u_{meas}^2 + u_{workingRM}^2} = \sqrt{u_{meas}^2 + u_{meas}^2 + u_{CRM}^2} = \sqrt{2u_{meas}^2 + u_{CRM}^2} \qquad (7.7)$$

where $u_c$ in equation 7.7 is approximately $\sqrt{2}\, u_{meas}$ if $u_{mean} \gg u_{CRM}$. The propagation of uncertainties can be ameliorated by making multiple measurements of the working standard. If the value of the quantity of the working standard is measured $n$ times and averaged, the uncertainty of the average is now

$$u_c = \sqrt{u_{meas}^2 + u_{workingRM}^2} = \sqrt{u_{meas}^2 + \frac{u_{meas}^2}{n} + u_{CRM}^2} = \sqrt{u_{meas}^2\left(1 + \frac{1}{n}\right) + u_{CRM}^2} \qquad (7.8)$$

which, as $n$ becomes large, tends to the uncertainty of equation 7.6.

The moral of this story is that the more you are willing to work at establishing the quantity value of your working standard, the less the penalty in measurement uncertainty you will have to pay.

## 7.4 Examples of Metrological Traceability Chains

### 7.4.1 Breathalyzer

In Australia there have been discussions as to how to make evidential breathalyzer measurements traceable to national or international standards. The amount of ethanol in a motorist is required by law to be less than a prescribed concentration, for many countries between 0.0 and 0.08 g per 100 mL blood. Recently there has been a move to change legislation to a limit of a given mass per 210 L of breath to avoid arguments about the blood/breath partition coefficient. As with any forensic measurement, metrological traceability is a key component of a successful prosecution. At present police buy standard solutions of ethanol in water from certifying authorities, with which they calibrate the evidential breathalyzers before use. These breatha-lyzers might be based on infrared measurements or electrochemical oxida-tion of the ethanol. The companies that sell ethanol standards establish the concentration of the ethanol by gas chromatography or by titration with potassium dichromate solution. In Australia it has been proposed that the National Measurement Institute certify a standard ethanol solution by a primary method (isotope dilution mass spectrometry, IDMS) and use this to certify, in turn, pure potassium dichromate, which can be sold to the certi-fying authorities, or other ethanol standards for GC measurements. A full traceability chain for a measurement made with calibration of standards by GC is given in figure 7.8.

The top of the chain is the SI unit of amount-of-substance concentration mol dm$^{-3}$ (mol L$^{-1}$). A certificate from the National Measurement Institute gives confidence to the certifying authority that the amount-of-substance

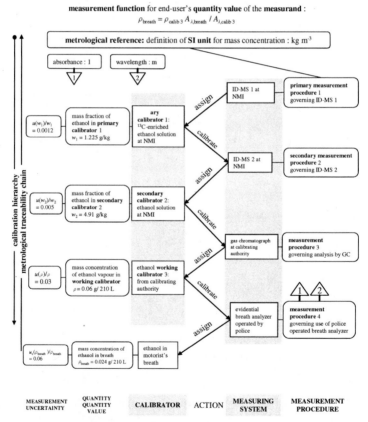

**Figure 7.8.** A possible metrological traceability chain for the result of a breathalyzer measurement of a motorist's breath alcohol.

concentration has been established with proper metrological traceability to the appropriate metrological reference. Down the chain, the calibration of ethanol standards passes on the traceability with an increased measurement uncertainty.

### 7.4.2  Protein in Grain

The protein content of grain is used to establish the price the commodity will fetch. This is an example of an industry-accepted reference method that still needs appropriate standards. The nitrogen content of a sample of grain is measured by near infrared (NIR) spectroscopy, and this is multiplied by a conventional factor of 6.25 to give the protein content. This is a conventional procedure, but standard samples of grain are still required to calibrate the measuring system (the NIR apparatus and attendant equipment). The nitrogen content of a standard grain is established by calibration using a pure organic reference material, examples are 2-amino-2-hydroxymethyl-1,3-propanediol (tris) and ethylene diamine tetraacetic acid (EDTA). A possible chain in which tris, obtained as a certified reference material from the National Institute of Standards and Technology (NIST) is the primary calibrator is shown in figure 7.9.

This chain is a good example of a prescribed measurement procedure (the use of NIR for nitrogen and a specified conversion factor) giving the value of measurand. Although the protein content of the standard grain sample is in SI units through calibration using the organic compound with SI-traceable nitrogen content, the measurand must be described as "the mass fraction of protein in grain measured by NIR as 6.25 × [N]," with [N] clearly defined. If another laboratory analyzed the grain by a method that measured the protein content directly, the result would be different from the one obtained by the nitrogen content method. It is therefore important to clearly state the measurand for measurements that rely on defined procedures.

## 7.5  Checklists for Metrological Traceability

Before embarking on a chemical measurement, some thought must be given to the measurand (exactly what quantity is intended to be measured), the requirements for measurement uncertainty, metrological traceability, and the measurement method. Often the choice of method is prescribed or obvious, but it is important not to complete a complex measurement only to realize that the measurement uncertainty is too large for the results to be useable or that a suitable calibrator was not available and so metrological traceability cannot be claimed. The process can be iterative, as occasionally the dictates of metrological traceability or measurement uncertainty lead to changes in the method.

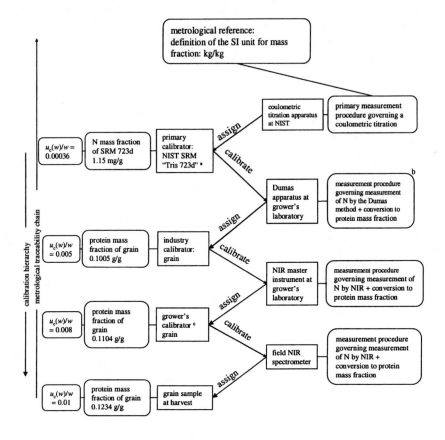

ª Tris = 2-amino-2-hydroxymethyl-1,3-propanediol
ᵇ Master instruments are calibrated and then certify grain samples in an interlaboratory study
ᶜ The master instruments each measure grain samples to act as grower's calibrator for field measurements

**Figure 7.9.** A possible metrological traceability chain for the result of a measurement of protein in a sample of grain. [a]Tris = 2-amino-2-hydroxymethyl-1,3-propanediol; [b] Dumas apparatus is calibrated using tris CRM and grain samples are certified in an interlaboratory study; [c]the master instruments measure grain samples to act as the grower's calibrator for field measurements.

### 7.5.1 A Metrologically Sound Checklist

This checklist for metrological traceability is a subset of a possible checklist for measurement, perhaps without concerns about method validation:

- Define the measurand and choose the measurement method.
- Identify requirements for metrological traceability. In a very simple system, a single metrological traceability chain may lead to a single

reference. However, when defining the scope of the analysis, procedures for corrections of recovery or bias quantity values carried by reference materials also need to be traceable. Decide to what extent influence factors such as temperature, need to be traceable.

- Select metrological reference(s). Metrological traceability can only be established to an existing and documented metrological reference. In most cases the reference will be the definition of the measurement unit of the measurement result.
- Select calibration hierarchy. By selecting a working calibrator, its calibration hierarchy is determined by the available documentation. Attention should also be paid to the calibration and metrological traceability of measurement results for input quantities to a measurement function measured by accessory equipment such as balances, thermometers, and volumetric ware.
- Acquire and verify a certified calibrator (CRM) from which a working calibrator is to be prepared. A working calibrator should be verified for integrity, validated for commutability, and have documented metrological traceability.
- Perform measurement.
- Document metrological traceability. This requires identification of all CRMs used as calibrators, calibration certificates for equipment, and a statement of the measurement uncertainty of the measurement result. The metrological traceability chain is thus established.
- Report metrological traceability. Measurement reports may require details of the metrological traceability chain or at least a statement of the metrological reference.

### 7.5.2  A Practical Checklist

If the analyst in a field laboratory can identify a CRM for calibration of working calibration solutions, and its certificate gives sufficient evidence that the CRM embodies a metrologically traceable quantity value, then his or her work is just about done (figure 7.10). The intricacies of what national measurement institutes and calibration laboratories did to ensure that the CRM has metrologically traceable quantity values are all paid for in the certificate.

Having understood the complete requirements for metrological traceability, a practical checklist can be written:

- Define measurand and choose measurement method.
- Choose and acquire calibrator that has documented metrological traceability. Assess metrological traceability requirements of mass, volume and temperature results.
- Prepare any working calibrators and estimate measurement uncertainty of the quantity value of the calibrator.
- Perform measurement.
- Estimate measurement uncertainty of measurement result.
- Report measurement result and document metrological traceability.

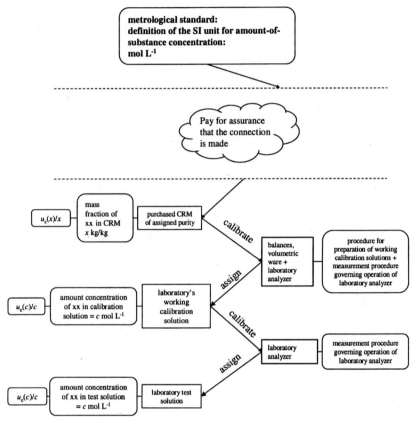

**Figure 7.10.** A metrological traceability chain from the point of view of an end user.

### 7.5.3   Reality Traceability

Laboratory personnel commonly give one of three common responses to someone who points out the traceability requirements of ISO/IEC 17025: "Its ok; we have our balances calibrated every year," and when it is further pointed out that calibration extends to the chemical aspects of the measurement, "I am sure it is ok to use this analytical-reagent-grade material to make a calibration solution," or "we make measurements in such complex matrices, a suitable CRM just does not exist." The first comment stems from the rather physical or engineering flavor of the section in ISO/IEC 17025 (section 5.6) (ISO/IEC 2005), although traceability of calibration materials is mentioned.

The second response about the source of the calibration material is more difficult to deal with, because the truth is that the measurements being made are almost certainly not metrologically traceable. Faith in a percentage purity on the label of a bottle might please the manufacturer, but it does nothing for genuine confidence in the results of a measurement performed using this material as a calibrator. The answer is to bite the financial bullet and buy a CRM, prepare working calibration standards, and make sure the measurement uncertainty is properly taken into account.

Regarding the last common response, there are two approaches to deal with the metrological traceability of a result from a complex matrix, such as is found in environmental and biological samples. If the measurement procedure involves separation of the analyte from the matrix, with the calibrated instrumental measurement being a pure material, (e.g., where chromatography has been used to separate and identify the analyte), then the assignment of a quantity value in this step can be performed after calibration by a pure reference material. This leaves the metrological traceability of any sampling and sample preparation steps to be established, which is usually accomplished when measuring the recovery. If a CRM is used in a spiking experiment, then metrological traceability of this part of the overall analysis is to the metrological reference of the CRM. Of course, considerations such as commutability of the CRM spike and its measurement uncertainty must also be addressed. If a measurement is made directly on the sample, modern biosensors are being developed for use with whole blood, for example, then metrological traceability will still be through whatever calibration material is used. So if a pure CRM is added as a spike in a standard addition experiment, then the metrological traceability is to the metrological reference of the quantity value of this material. The argument advanced by EURACHEM (EURACHEM and CITAC 2002) is that it might not be best practice to calibrate with a matrix CRM, but to use it instead to estimate recovery and calibrate with a pure material. EURACHEM states that is generally not possible to completely match the properties of the matrix CRM with those of the sample,—the cost could be prohibitive and that the measurement uncertainty might become too great. (If a correction for recovery is made, then this is likely to be the case anyway.)

Perhaps if you have a resourceful and supportive national measurement institute, you might be able to persuade it to organize an interlaboratory certification to provide appropriate calibration materials, as has been done in Australia by the grain industry for protein measurements.

## Notes

1. The author is a task group member of an ongoing IUPAC project 2001-010-3-500 titled "Metrological traceability of measurement results in chemistry," http://www.iupac.org/projects/2001/2001-010-3-500.html. The author has drawn heavily on the work of the project, but takes full responsibility for the contents of this chapter.

## References

Becker, P (2003), Tracing the definition of the kilogram to the Avogadro constant using a silicon single crystal. *Metrologia*, 40 (6), 366–75.

BIPM (2006), The international system of units. Available: http://www.bipm.fr/en/si/ (Sèvres, International Bureau of Weights and Measures).

Emanuele, G, et al. (2002), Scripps reference gas calibration standards system: for carbon dioxide-in-nitrogen and carbon dioxide-in-air, revision of 1999 (La Jolla, CA: Scripps Institution of Oceanography).

EURACHEM (2000), *Quantifying uncertainty in analytical measurement*, 2nd ed., M Rosslein, S L R Ellison, and A Williams (eds.) (EURACHEM Teddington UK).

EURACHEM and CITAC (2002), *Traceability in chemical measurement, EURACHEM/CITAC guide*, 1st ed. (EURACHEM/CITAC Teddington UK).

ISO (1993), *International vocabulary of basic and general terms in metrology*, 2nd ed. (Geneva: International Organization for Standardization).

ISO/IEC (2005), General requirements for the competence of calibration and testing laboratories, 17025 2nd ed. (Geneva: International Organization for Standardization).

Joint Committee for Guides in Metrology (2007), *International vocabulary of basic and general terms in metrology*, 3rd ed. (Geneva: International Organization for Standardization).

NPL (National Physical Laboratory) (2005), The Avogadro Project. Available: http://www.npl.co.uk/mass/avogadro.html.

# 8

## Method Validation

### 8.1 Introduction

Many aspects of a chemical analysis must be scrutinized to ensure that the product, a report containing the results of the analysis, fulfills the expectations of the client. One of the more fundamental factors is the analytical method itself. How was it chosen? Where does it come from? When a laboratory is faced with a problem requiring chemical analysis, there may be set methods described in a standard operating procedure, but often the analyst might have to make a choice among methods. For the majority of analytical methods used in field laboratories, there is neither the expertise nor the inclination to start from scratch and reinvent the wheel. The analyst wants a method that can be implemented in his or her laboratory. Compilations of methods that have been evaluated do exist and have the imprimatur of international organizations such as the International Organization for Standardization (ISO) or the American Society for Testing and Materials (ASTM). Failing this, the scientific literature abounds in potential methods that have the recommendation of the authors, but may not always be as suitable as claimed.

This chapter has two aims: to demonstrate the necessity of using properly validated and verified methods and to explain what constitutes a validated method, and to provide an introduction to method validation for in-house methods. There is an abundance of published material that defines, describes, and generally assists with method validation, some of which is referenced here (Burgess 2000; Christensen et al. 1995; EURACHEM 1998; Fajgelj and Ambrus 2000; Green 1996; Hibbert 2005; ICH 1995, 1996; LGC 2003; Thompson et al. 2002; USP 1999; Wood 1999).

## 8.2 Method Validation—Some Definitions

"Method validation" is a term used for the suite of procedures to which an analytical method is subjected to provide objective evidence that the method, if used in the manner specified, will produce results that conform to the statement of the method validation parameters. Like many aspects quality assurance, method validation is of a relative nature. As with the concept of fitness for purpose, a method is validated for a particular use under particular circumstances. If those circumstances vary, then the method would need to be re-validated at least for the differences. Common sense should be used, and the analysts should use his or her skill and experience to decide what aspects of a method require validation and to what extent. The goal of satisfying client requirements is prominent in most published definitions of method validation, some of which are listed below:

- Method validation involves the evaluation of the fitness of analytical methods for their purpose
- The process of proving that an analytical method is acceptable for its intended purpose (Green 1996).

The ISO defined method validation as (ISO 1994a):

1. The process of establishing the performance characteristics and limitations of a method and the identification of the influences which may change these characteristics and to what extent. Which analytes can it determine in which matrices in the presence of which interferences? Within these conditions what levels of precision and accuracy can be achieved?
2. The process of verifying that a method is fit for purpose, i.e. for use for solving a particular analytical problem.

In a later revision (ISO 2005, term 3.8.5) the definition became: Confirmation, through the provision of objective evidence, that the requirements for a specific intended use or application have been fulfilled.

ISO/IEC 17025 (ISO/IEC 2005, section 5.4) states that method validation is "confirmation by examination and provision of objective evidence that the particular requirements for a specified intended use are fulfilled," and it stipulates protocols for validating in-house methods. Proper documentation is proof of validity. It is unlikely that many interested parties were present during the experiments and subsequent data analysis that demonstrated validity. Therefore, in many ways method validation is only as good as its documentation, which describes the method, the parameters investigated during the method validation study, the results of the study, and the conclusions concerning the validation. The documentation of the validity of a method provided by a laboratory seeking accreditation to an international standard must pass the scrutiny of an assessor.

In some sectors, particularly food and health, the requirement for fully validated methods is prescribed in legislation, and the starting point for any method validation should be the standards and protocols emanating from the controlling organization. For example, the Codex Alimentarius contains such directives, and the United Kingdom drinking water inspectorate guidelines state: "A laboratory using an analytical method which is not referenced to a fully validated authoritative method will be expected to demonstrate that the method has been fully documented and tested to the standard currently expected of an authoritative reference method" (DWI 1993, paragraph 13).

## 8.3  When Should Methods Be Validated?

No method should be used without validation. The major part of the work of validation might have been done elsewhere at an earlier time, but even then, the analyst should be satisfied that the method as used in his or her laboratory is within the validation specifications. This will be discussed below. ISO/IEC 17025 encourages the use of methods published in international, regional, or national standards and implies that if these methods are used without deviation, then the validation requirements have been satisfied. What does need to be validated are nonstandard methods, methods designed and developed by individual laboratories, standard methods used outside their scope, and amplifications and modifications of standard methods.

### 8.3.1  The Development and Validation Cycle

If a method must be developed from scratch, or if an established method is changed radically from its original published form, then before the method is validated, the main task is simply to get the method to work. This means that the analyst is sure that the method can be used to yield results with acceptable trueness and measurement uncertainty (accuracy). When the analyst is satisfied that the method does work, then the essentials of method validation will also have been done, and now just need to be documented. If there is an aspect of the method that does not meet requirements, then further development will needed. Discovering and documenting that the method now does satisfy all requirements is the culmination of method validation.

The foregoing discussion is encapsulated in figure 8.1, which shows that method validation can be seen as checking and signing off on a method development cycle, with the important caveat that initial development is often done in a single laboratory and so will not usually establish interlaboratory

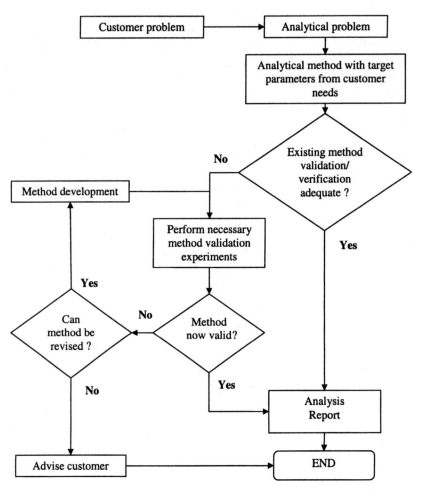

**Figure 8.1.** The relationship between validation, development, and customer requirements. (Adapted from Hibbert 2005.)

reproducibility. There has to be a final assessment of measurement uncertainty to make sure all appropriate aspects of the method have been accounted for.

### 8.3.2 Interlaboratory Studies

An important aspect of a full method validation is estimating bias components attributable to the method itself and to the laboratory carrying out the analysis. This step is required to estimate measurement uncertainty with a reasonable range that covers results that would be obtained in another laboratory of similar experience and standing. In chapter 5 I discussed these approaches at length. ISO (1994b) has a procedure for such interlaboratory studies, and Horwitz (1995) has published International Harmonized Protocols. If the validation is only done in house, reproducibility standard deviation must be accounted for by using, for example, the Horwitz formula, $\sigma_H = 0.02\, f^{0.8495}$, where $f$ is the value of the measurand expressed as a mass fraction, and $\sigma_H$ is the reproducibility standard deviation also expressed as a mass fraction. For concentrations > 120 ppb, this value is usually within a factor of 2 of results obtained from interlaboratory studies.

### 8.3.3  Verification

Much of the work of method validation is done by international organizations that publish standard methods. The reason such methods appear to be written in a kind of legalese is that there must be no doubt as to what the method is and how it should be implemented. When accuracy and precision data are published from interlaboratory trials, there is some confidence that the method has undergone extreme scrutiny and testing. A laboratory that uses a method for the first time should spend some time in going through the analysis with standard materials so that when used with field samples, the method will yield satisfactory results. This is verification and must be done to an appropriate level before any method is used. By its nature, verification comes under the heading of Single Laboratory Validation (Thompson et al. 2002). A minimum set of verification experiments is given in table 8.1.

Another circumstance in which verification checks must be performed is when a particular aspect of the method or its implementation is changed. Common changes are new analyst, new equipment or equipment part (e.g.,

Table 8.1.  Minimum experiments that should be performed to verify the use of a standard method before first use in a laboratory

| Parameter | Verification experiments |
|---|---|
| Bias/recovery | Analysis of a matrix-matched reference material, or typical matrix spiked with reference material, or typical test material with quantity value with known measurement uncertainty established by another method |
| Precision | Standard deviation of ten independent determinations of typical test materials |
| Measurement uncertainty | Consideration of factors that could contribute to measurement uncertainty from the use in the analysts laboratory; revision of the uncertainty budget accordingly |
| Calibration model (linearity) | When calibrating the method for the first time attention must be given to whether the calibration equation is properly followed |
| Limit of detection (LOD) | If necessary; analysis of test material near the LOD to demonstrate capability of detection at that level |

new column, new detector), new batch of reagent that is subject to batch variation (e.g., enzymes, antibodies) or changes in the laboratory premises. A minimum verification is to analyze a material before and after the change and check for consistency of the results, both in terms of mean and standard deviation. (Following such changes using a CuSum control chart is discussed in chapter 4.)

## 8.4 Parameters Studied in Method Validation

A tradition has grown up in method validation in which a sequence of tests taken together result in a suitably validated method. Because of the age of the approach, the relationship of measurement uncertainty and metrological traceability of a result to method validation has not always been clear. Establishing an acceptable measurement uncertainty of a result is one of the major planks of method validation, and discussions of precision and accuracy are found under this heading. Because some analytical systems these days have multivariate or nonlinear calibration, the traditional terminology of "linearity" and "linear range" have been changed to reflect the wider meaning of suitability of the calibration model. Table 8.2 gives the performance parameters described below, together with their more traditional names and groupings.

### 8.4.1   Horses for Courses

Not all methods require each parameter detailed in table 8.2 to be established. For example, a method that only measures the active ingredient in a 100-mg cold cure as part of a quality control protocol is not concerned with limit of detection, the matrix is fixed, and the calibration range might only need to be established between 80 and 120 mg. An analysis that determines the presence or absence of the target analyte needs only to establish its selectivity, limit of detection, and ruggedness. Table 8.3 details some common analytical systems with their critical method validation parameters.

Measurement uncertainty is a critical parameter for nearly every kind of analytical system. Parameters in the second column of table 8.3 are not unimportant and must be established, but they are not likely to become limiting factors in the development of the method. In table 8.3, where selectivity is in parentheses, this is not to say that the method should not be demonstrably capable of analyzing the target analyte, but that it should be clear very quickly whether or not the method is doing its job.

### 8.4.2   How Much Validation?

The fit-for-purpose nature of chemical analysis tells us that there is an appropriate level of method validation. Too much and resources are squandered for no benefit; too little and the product could be worthless. Modern

Table 8.2.  Method performance parameters assessed in a method validation study

| Parameter[a] | Description and comments |
|---|---|
| Identity | Measurement correctly applies to the stated measurand |
| Selectivity<br>Specificity | Determination of the extent of the effects of interfering substances and the ability of the method to measure the measurand; analysis in different matrices covered by the scope of the validation |
| Limits | |
| Limit of detection | Minimum value of the measurand at which the presence of the analyte can be determined with a given probability of a false negative, and in the absence of the analyte, at a given probability of a false positive |
| [Limit of determination] | Minimum value that can be obtained with a specified measurement uncertainty |
| Calibration [linearity] | |
| Model parameters [sensitivity] | Adequacy of the calibration model; parameters with uncertainties. |
| Calibration range [linear range] | Range of values of the measurand in which the validation holds |
| Bias and recovery [accuracy] | Demonstration of the absence of significant systematic error after corrections have been made for bias and/or recovery |
| Measurement uncertainty | |
| Type A effects [repeatability and reproducibility precision] | Uncertainty estimated from statistical treatment of the results of repeated analyses |
| Type B effects | Estimates based on nonstatistical methods including published data and experience of the analyst |
| Robustness or ruggedness | Ability of method to remain unaffected by small variations in method parameters (some authors make the distinction be-between the property robustness and a ruggedness test in which deliberate changes are made in a method to assess the robustness) |

[a]Parameters in square brackets are headings used in a classical approach to method validation.

practice puts great emphasis on the dialogue with the client and working through to fulfill the needs of the client. But the majority of clients do not know what they really want in terms of method validation, or any other quality parameter, for that matter. The professional analyst is charging as much for his or her accumulated expertise as they are for the time and

Table 8.3. Different analytical systems with their method validation requirements

| Analytical system | Critical method validation parameters | Other validation parameters[a] |
|---|---|---|
| Qualitative analysis only | Selectivity, LOD | Ruggedness |
| Identification and analysis of a particular form of a substance (e.g., oxidation state, isomer, conformer) | Selectivity | Calibration, MU, ruggedness, (LOD) |
| Measurement of an analyte in a matrix | Selectivity, MU | Calibration, ruggedness, (LOD) |
| Trace analysis | Selectivity, MU, LOD | Calibration, ruggedness, (LOD) |
| Results that will be compared with those from other laboratories | MU | Calibration, ruggedness, (LOD) |
| Results that will be compared with limits or specifications | MU | Ruggedness, (LOD, selectivity) |
| Method that is implemented by different analysts on different instruments | MU, ruggedness | Calibration, (LOD, selectivity) |
| Require to analyze to target measurement uncertainty | MU | Calibration, ruggedness, (LOD) |
| A number of samples having a range of concentrations | Calibration, MU | Ruggedness, (LOD) |

[a]LOD = limit of detection, MU = measurement uncertainty. Parameters in parentheses need only be studied if required.

analyses being performed. So the onus can well fall back on the analyst to make an informed decision in the interests of the client. Accreditation organizations and other legal watchdogs are there to make sure in general terms that the laboratory is doing the right thing. Here is a continuum ranging from a new method to be specified as a standard method in a highly regulated industry, in which an international body conducts lengthy interlaboratory studies followed by further validation or verification in the laboratory in which the method will be used, to an ad hoc method that will only be used once and for which the results are not of such importance. In the latter case there is still no point in using a method in which there is no confidence that the results are suitable, but the analyst may decide that the experiment is still worth doing. In fact, an analyst should assess whether a measurement will give useful results before any measurement actually takes place.

In the majority of cases in professional analytical laboratories, a standard method is the starting point for a routine laboratory method, and many of the validation parameters are taken from the published description of the method. The laboratory then must spend some time verifying and validating necessary amendments to the method. When the client is involved in the decision of what aspects to validate and to what depth to validate, it is a good idea to include the agreed-upon protocols in the contract, so there can be no argument later if it turns out that a vital aspect was omitted.

The level of validation to be undertaken must be chosen considering scientific and economic constraints. All data have some value, and results from the development phase can all be pressed into service for validation. Separate planned experiments might lead to better and more statistically defensible results, but when this cannot be done, then whatever data are at hand must be used. The best use can be made of experiments to be done by understanding what is required. For example, in a precision study, if the goal is to know the day-to-day variability of an analysis, then duplicate measurements over 5 days would give more useful information than 5 replicates on day 1, and another 5 on day 5. The strategy would be reversed if variations within a day were expected to be greater than between days.

## 8.5 Specifying the Method and Target Validation Parameters

Before an experiment is done, there is an important phase in which the road map to validation is made. First the method itself is clearly specified in terms of a measurement procedure based on a principle of measurement, including equations for any calculations necessary to obtain a result. Then the parameters that need to be validated are determined. In the light of the expected use of the method, and possibly with the input of the client, critical parameters are determined that will show whether the validated method is fit for its purpose. A target measurement uncertainty will usually be specified, particularly if the measurement result will be used to compare with a legal limit or product specification. Although the world is moving toward reporting the measurement uncertainty associated with a measurement result, the concept may be expressed in terms of a maximum bias (tolerance) and repeatability or reproducibility precision (as a relative standard deviation or coefficient of variation). In addition, a target limit of detection can be stated. Particular influence factors may be identified and their effects prescribed. For example, the temperature range in which the target measurement uncertainty must hold may be specified, or potential interferences that should not affect the method may be specified. The report of the validation will then address each of the parameters and conditions and demonstrate that the method successfully complies with the values and ranges. If it does not, then the options shown in figure 8.1 are followed. Either the method is further developed and revalidated, or the conditions for validation are relaxed, or the project is abandoned.

It may be that the method as developed has no particular client in mind. Now the method validation process is to discover and report the values of the parameters for the method as specified. In the future a prospective user of the method can judge whether it will be suitable for his or her purpose, and perhaps cause a further round of the development and validation cycle. The description of a standard method with information about repeatability, reproducibility, and so on, is a kind of validation report, although it would be useful if standardization bodies actually issued a method validation document together with the description of the method.

## 8.6 Obtaining Method Validation Parameters

### 8.6.1  Identity

Confirming that the analytical method does indeed result in the value of the measurand, on one hand, might be as trivial as an assertion to that effect, because the nature of the method cannot fail to deliver it, or, on the other hand, identifying a given form of a molecule in a complex matrix could be the most time consuming and costly part of method validation. Having heard defense attorneys asking whether the signal allegedly attributed to an illegal substance could possibly have arisen from something more benign, I know that ultimately there could be a substance in the test portion that "walks like a duck and quacks like a duck" but is not a duck. There has been some work on the uncertainty of qualitative analysis based on a Bayesian approach (Ellison et al 1998), but analysts eventually use their judgment to assert that a method is specific. When in doubt, the best solution is to conduct the analysis using methods based on different physical or chemical principles to confirm identity. For example, infra-red or NMR spectroscopies could be used to confirm GCMS data, or libraries of spectra can be used to add confidence to the assignment of structure. When there is concern about a particular source of misidentification, then specific experiments should be performed to check that the correct isomer or species is being analyzed. If there is the possibility of *cis-trans* isomerism about a double bond, for example, it is often possible to predict the $^1$H NMR coupling from each isomer. This discussion highlights the need to specify the measurand carefully in the method protocol. There may be a great difference in the concentration of copper, inorganic copper, and bioavailable copper and in the ability of a method to correctly identify the particular species of interest (see discussions in chapter 10).

#### 8.6.1.1  Selectivity and Specificity

Selectivity is a measure of the extent to which the result gives the value of the measurand and only of the measurand. If a method is 100% selective, it is said to be *specific* to the measurand. There are many ways of ensuring selectivity, and it is a basic characteristic of the analytical method. Chromatography, in general "separation science," is a powerful analytical tool because of its ability to present individual species to the detector. Chiral separations are the ultimate in discriminating among species, in this case optical isomers. When it is impossible to effect a separation, the analysis must take account of interferences in some other way. It may be that an interfering species of similar concentration to the analyte does not produce as great a signal from the instrument. Interference studies can quantify the effects of nominated interfering substances, and either they can be deemed insignificant or the presence of the interferent may be treated as a systematic error and corrected for or allowed for in the measurement uncertainty. (A clever way of accounting for electroactive interferents in a channel biosensor is described by Zhao et al. [2003, 2004].) These procedures then limit the method to a particular matrix containing a particular set of interfering species or require a series of experiments to quantify the effect of the interferents in all possible test samples. This should be treated as an extension of the method and must also be validated, but it could be a useful solution to the problem of interferences. Modern chemometric techniques that rely on multivariate data are often used to make a method specific.

With chromatography it must still be demonstrated that apparently single peaks do truly arise from the single compound of interest. The advent of methods and detectors that add a second dimension provide this surety. Two-dimensional GC, also known as comprehensive GC or GC × GC, is such a technique in which, by modulation of the temperature of the output of a first column, species separated on this column can be further processed on a shorter, second column operating on a different principle of separation (Ong and Marriott 2002). This allows complex mixtures, such as petroleum samples, to be better separated. Chromatography with mass spectrometry can also demonstrate purity of a peak, if the mass spectrum of the compound eluting is known. Diode array detectors operating at a number of wavelengths as detectors in liquid chromatography may also reveal multiple components.

### 8.6.1.2  Interference Studies

How analytical methods deal with interferences is one of the more ad hoc aspects of method validation. There is a variety of approaches to studying interference, from adding arbitrary amounts of a single interferent in the absence of the analyte to establish the response of the instrument to that species, to multivariate methods in which several interferents are added in a statistical protocol to reveal both main and interaction effects. The first question that needs to be answered is to what extent interferences are expected and how likely they are to affect the measurement. In testing blood for glucose by an enzyme electrode, other electroactive species that may be present are ascorbic acid (vitamin C), uric acid, and paracetamol (if this drug has been taken). However, electroactive metals (e.g., copper and silver) are unlikely to be present in blood in great quantities. Potentiometric membrane electrode sensors (ion selective electrodes), of which the pH electrode is the most famous, have problems with similar ions interfering. For example, hydroxyl ions interfere with a fluoride-ion-selective electrode. In a validation this effect must be quantified and included in the measurement uncertainty, together with a statement of the pH range in which results will be obtained that conform to the stated uncertainty. There are standard and recommended methods for quantifying interferences; for ion-selective electrodes, see Buck and Lindner (1994).

### 8.6.2   Limit of Detection

Limit of detection (LOD) sounds like a term that is easily defined and measured. It presumably is the smallest concentration of analyte that can be determined to be actually present, even if the quantification has large uncertainty. The problem is the need to balance false positives (concluding the analyte is present, when it is not) and false negatives (concluding the analyte is absent, when it is really present). The International Union of Pure and Applied Chemistry (IUPAC) and ISO both shy away from the words "limit of detection," arguing that this term implies a clearly defined cutoff above which the analyte is measured and below which it is not. The IUPAC and ISO prefer "minimum detectable (true) value" and "minimum detectable value of the net state variable," which in analytical chemistry would become "minimum detectable net concentration." Note that the LOD will depend on the matrix and therefore must be validated for any matrices likely to be encountered in the use of the method. These will, of course, be described in the method validation document.

Consider a measurement made of a blank—a sample that is identical to a routine sample for measurement but without any analyte. A large number of measurements will lead to a distribution of (nonzero) results, which might or might not be normally distributed. A normal distribution is shown in figure 8.2, although this begs the question of what to do about apparently negative results that might be obtained when making measurements of concentrations near zero.

A critical limit may be defined at which the probability of making the error of rejecting the hypothesis that the measurement comes from a sample having zero analyte (a Type I error) is, say, 0.05. This cannot be taken as the LOD because, although the probability of a false positive is 0.05, and quite acceptable, the probability of finding a false negative is exactly 0.5; that is, half of all measurements made of an analyte having a concentration $L_{crit}$ will be less than $L_{crit}$ and rejected (figure 8.3). A sensible approach is therefore to shift the decision limit, the LOD, up in concentration until the probabilities of making Type I and Type II errors at $L_{crit}$ are equal (see figure 8.4).

It has been suggested that the LOD should be the concentration for which the lower limit of the confidence interval just reaches zero, and while this has a logic to it, as the value is not known until measurements are made, the analyst could be in for a number of speculative experiments before finding the magic concentration that satisfies the condition.

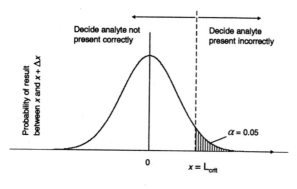

Concentration (x)

**Figure 8.2.** The distribution of blank measurements about zero. $L_{crit}$ is a concentration at which the probability of making a Type I error (deciding analyte is present when it is not, $\alpha$) is 0.05.

### 8.6.2.1 The Three-Sigma Approach

A straightforward and widely accepted approach is to deem that an instrument response greater than the blank signal plus three times the standard deviation of the blank signal indicates the presence of the analyte. This is consistent with the approach shown in figure 8.4 if it is assumed that the standard deviation of a measurement result at the LOD is the same as that of a blank measurement. Suppose there is a linear calibration relation

$$y_i = a + bx_i + \varepsilon \tag{8.1}$$

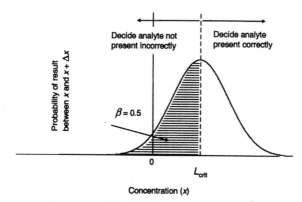

Concentration (x)

**Figure 8.3.** Distributions of measurements of analyte of concentration, $L_{crit}$, showing that half the results at this concentration will be rejected using $L_{crit}$ as the decision point.

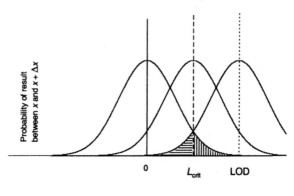

Concentration (x)

**Figure 8.4.** Choice of a limit of detection (LOD) at which the probabilities of Type I and Type II errors are equal at $L_{crit}$. The distributions are, from left to right, measurements of a blank with analyte concentration zero, measurements of a sample of concentration $L_{crit}$, and measurements of a sample of concentration LOD.

where $y_i$ is the indication of the measuring instrument for a particular value of the concentration of the analyte $(x_i)$, $a$ is the intercept, $b$ is the slope of the calibration line, and $\varepsilon$ is the random component of the measurement with expectation zero and standard deviation $\sigma$. A blank reading $\Delta y_b + 3\sigma_b$ corresponds to a concentration of

$$x_{DL} = \frac{\Delta y_b + 3\sigma_b - a}{b} \tag{8.2}$$

The measurements that should be made are 10 independent determinations of a blank solution. The sample standard deviation of these 10 measurements is taken as an acceptable estimate of $\sigma_b$. For methods that do not give a response in the absence of analyte (for example, many ion-selective electrodes do not give a stable signal in the absence of the target ion or a known interfering ion), a blank is spiked at the lowest acceptable concentration. Again, 10 measurements of independently prepared test solutions are made, and the sample standard deviation is taken as the estimate of $\sigma_b$. It can be shown that the "blank plus three sigma" gives a probability of both false positives and false negatives of about 7%. If the limit is to be 5% errors (a 95% confidence), then sigma is multiplied by 3.29. This is recommended by IUPAC (Currie 1995).

### 8.6.2.2   LOD Estimation in the Absence
of Blank Measurements

If the measurement of LOD is not critical, an estimate can be made from the calibration parameters taking the intercept as the blank measurement and the standard error of the regression as the standard deviation of the blank. Equation (8.1) becomes

$$x_{DL} = \frac{3s_{y/x}}{b} \tag{8.3}$$

Equation 8.3 relies on the assumption that the calibration parameters hold to the limit of detection, which is not necessarily correct. If the calibration is taken over a very wide range, for example in some element analyses by inductively coupled plasma, the uncertainty is not constant but is proportional to the concentration. In this case it is not possible to estimate the detection limit by equation (8.3).

### 8.6.2.3   A More Sophisticated Approach Based on
Calibration

IUPAC recommends the following formula for calculating the limit of detection:

$$x_{DL} = \frac{2t_{0.05'n-2}s_{y/x}}{b} \sqrt{\frac{1}{m} + \frac{1}{n} + \frac{\bar{x}^2}{\sum_i (x_i - \bar{x})^2}} \tag{8.4}$$

where $t_{0.05',n-2}$ is the one-tailed Student's $t$ value for $\alpha = 0.05$ and $n-2$ degrees of freedom, $m$ is the number of replicates of the measurement at the LOD, $n$ is the number of points in the calibration line, $\bar{x}$ is the mean value of the calibration concentrations ($x_i$), and $b$ is the slope of the calibration equation (see also Long and Winefordner 1983).

### 8.6.2.4   LOD for a Qualitative Method

If the analytical method is required only to detect the presence of the analyte but not report a measurement result, then there will be no calibration relation to go from an indication of the measuring instrument to a concentration. In this case a number of independent measurements (at least 10) must be made at concentrations in a range that brackets the expected LOD. The fraction of positive and negative findings is reported for each concentration, and a concentration is then chosen as the LOD that gives an acceptable proportion of positive results.

### 8.6.3   Calibration Parameters

A discussion about calibration must also include consideration of single-point calibration and direct comparison of responses to samples of known and unknown quantities. In each case the linearity of the calibration (i.e., the correctness of taking a ratio of instrument responses) is accepted in routine work. In method validation this assumption must be verified by making a series of measurements in a concentration range near to the range used, and the linear model must be demonstrated to be correct.

In traditional method validation, assessment of the calibration has been discussed in terms of linear calibration models for univariate systems, with an emphasis on the range of concentrations that conform to a linear model (linearity and the linear range). With modern methods of analysis that may use nonlinear models or may be multivariate, it is better to look at the wider picture of calibration and decide what needs to be validated. Of course, if the analysis uses a method that does conform to a linear calibration model and is univariate, then describing the linearity and linear range is entirely appropriate. Below I describe the linear case, as this is still the most prevalent mode of calibration, but where different approaches are required this is indicated.

Remember that for calibrated methods the working range is in terms of the concentration presented to the measuring instrument that is being calibrated, which might not be that of the original sample after dilution or other preparative steps. In documenting the method validation, it is acceptable, having quoted the working range of the calibration of the instrument, to indicate what range of concentrations of the sample this range would come from, as long as the sample pretreatment steps are explained fully in the description of the method.

The validation also will give information about the extent and frequency of calibration that will be necessary during use. Having established the linearity of the calibration by the methods given below, the analyst may decide that a single-point calibration is sufficient for routine use in the laboratory.

#### 8.6.3.1   Adequacy of the Calibration
####                 Model (Linearity)

Method validation must demonstrate that the calibration model (i.e., the equation by which the instrument response is related to a known value of a standard) holds for the system under investigation, and over what range of concentrations.

What makes a good calibration? This must be specified, and there is no accepted single path, but after collecting response data at a number of points (see below), the following steps can be followed:

First test the data for outliers. This is not as straightforward as performing a Grubbs's test for replicates of a measurement on a single analyte. Because the linear model correlates data, tests such as Grubbs's or Dixon's Q are not statistically valid. The effects of an outlier on the calibration line are different depending on whether it is near the mean of the data or at the extremes. This is illustrated in figure 8.5. An outlier at the extreme of the working range may be a reflection that the model no longer holds, or it may be a genuine outlier caused by some unforeseen error. Such an outlier is said to have great *leverage*. Because least squares calibration minimizes the sum of the squares of the residuals, a point lying far from the line contributes relatively more to that sum (residuals of 1 and 10 add 1 and $10^2 = 100$ respectively to the sum). Outliers should be viewed with great suspicion and should not be discarded lightly, unless the reason for the aberration is obvious. If possible, the measurement should be repeated, and if the result does not change much,

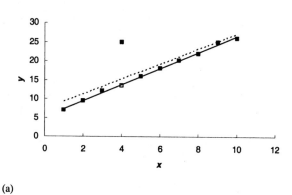

(a)

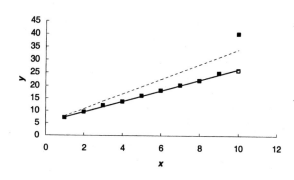

(b)

**Figure 8.5.** Outliers in calibration: (a) near the center of the data causing a shift in intercept and (b) at the extreme of the data (a point with high leverage), causing change in slope. The solid lines are fits to the data without outliers (solid points). The dashed lines are fits to the data with outliers (all points).

other measurements should be made in the vicinity of the outlier to eluci-
date what is causing the apparent deviation from linearity.

Then plot the calibration curve and residual plot. If the residuals are scattered nicely about zero with no particular trend or obvious outliers, then
the line is acceptable. A potential outlier is one that is more than twice the
standard error of the regression ($s_{y/x}$). (figure 8.6). Next assess the line for
curvature. If the real relation between instrument response and concentration is a curve, then this will be revealed in a residual plot, much more so
than might be apparent in the calibration curve. Curvature across the whole
range of concentrations invalidates the notion of a linear range, and at this
point a decision must be made about the calibration model. If the linear
model is inadequate for calibration, an equation that captures the relation
between instrument response and concentration should be used. This relation should be justified from knowledge of the chemical system. Merely
adding polynomial terms to the linear model until a better fit is discovered
or adding more principal components to a multivariate regression will always

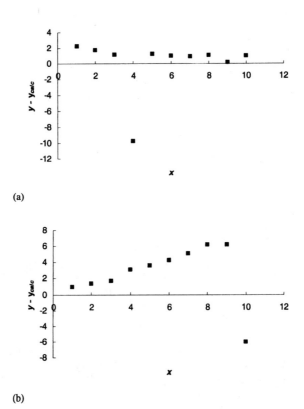

(a)

(b)

**Figure 8.6.** Residual plots of the calibration fits of
figure 8.5. (a) Outlier near center of data, (b)
outlier at extreme of data.

improve the calibration, but these practices run the risk of overfitting. Any such augmentation should be well studied and validated. A second fix is to reduce the linear range. Over a short enough span, a linear model can always be made to have an acceptably small error, but the question is, how short?

### 8.6.3.2  Working Range (Linear Range)

During the development of the method, a reasonable idea of the working range will be established, or there will be a target range set by the use of the method (in process control, for example). Many instrumental methods have a calibration curve that looks like the one depicted in figure 8.7. The concentration defining the limit of detection will be the absolute lower end, although the decision about where the working range starts must include some acceptable precision, and at the detection limit the uncertainty on any measurement may be considered too great. The concentration defining the start of the working range is sometimes known as the limit of determination. At the higher end of the concentration range, the instrumental response often saturates, leading to a flattening off of the calibration curve.

To establish linearity and the linear range, a blank solution, plus at least 6, but preferably 10, independent solutions should be prepared from a traceable reference material and presented to the instrument at least as duplicates and in random order. The range of these solutions then defines the linear range, assuming, of course, statistical analysis of the calibration data supports this contention. How is the range known without doing experiments? As stated above, there may be enough prior information to have a

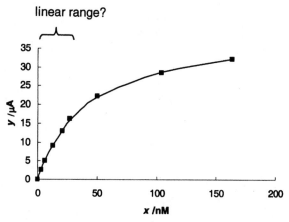

**Figure 8.7.** A response curve of an electrochemical measurement of copper with concentration of analyte. The line is a fit to the equation $y = y_0 Kx/(1 + Kx)$, $y$ is the measured current, $x$ is the concentration of copper, and $y_0$ and $K$ are constants of the model.

good idea of the range, and the use of the method may not be so stringent as to probe the real limits of the working range. However, it may be necessary to perform single measurements across the expected range, to make sure that the range is likely to be acceptable, before adding replicates.

If it is necessary to quantify adherence to the calibration equation, a method based on ANOVA can be used where replicate determinations at each point are made. Measurements are made on $k$ independently prepared solutions of different concentrations, where the $i$th concentration is measured $n_i$ times. These data are fitted to the calibration equation (this approach works for nonlinear calibrations as well). There are two reasons the $y$ values estimated from the calibration equation might not exactly agree with the measured $y$ values. First, there is the ubiquitous random measurement error, and second. discrepancies may result from the fact that the calibration equation does not actually fit (e.g., nonlinearity in a linear model). Replicate determinations of each calibration solution can only tell us about measurement error; the calibration has nothing to do with the variance of these measurements. How well the estimates from the calibration equation agree with the measured values depend on both measurement error and problems with the calibration equation. By comparing estimates of these variances (bad model + measurement error compared with measurement error), it should be possible to decide whether the model fits well enough. Using the nomenclature of ANOVA, the total sum of squares is given by

$$SS_T = \sum_{i=1}^{k}\sum_{j=1}^{n_i}(y_{i,j} - \bar{y})^2 \tag{8.5}$$

where $\bar{y}$ is the mean of all the responses of the instrument. $SS_T$ can be decomposed into three sums of squares, $SS_{ME}$, $SS_{LOF}$, and $SS_{REG}$, which are due to measurement error, lack of fit, and the regression, respectively:

$$SS_{ME} = \sum_{i=1}^{k}\sum_{j=1}^{n_i}(y_{i,j} - \bar{y}_i)^2 \tag{8.6}$$

$$SS_{LOF} = \sum_{i=1}^{k} n_i(\bar{y}_i - \hat{y}_i)^2 \tag{8.7}$$

$$SS_{REG} = \sum_{i=1}^{k} n_i(\hat{y}_i - \bar{y})^2 \tag{8.8}$$

where $\bar{y}_i$ is the mean of the replicate measurements on the $i$th calibration solution, and $\hat{y}_i$ is the value of $y$ of the $i$th calibration solution estimated by the regression equation. The better the data fit the calibration equation, the smaller $SS_{LOF}$, although this can never go to zero because the estimated value will always have a contribution from measurement error. As in ANOVA, the mean squares are the sum of squares divided by the degrees of freedom, $MS = SS/df$. $MS_{LOF} = SS_{LOF} / (k - 2)$ and $MS_{ME} = SS_{ME} / (N - k)$ ($N$ is the total number of data, and $k$ is the number of concentrations). $MS_{LOF}$ may be tested against $MS_{ME}$ using a one-sided $F$ test:

$$F = \frac{MS_{LOF}}{MS_{ME}}$$

(8.9)

The probability of $F$ is the probability of finding the observed ratio of mean squares given the null hypothesis that the data do indeed fit the calibration model. If this probability falls below an acceptable level (say, $\alpha = 0.05$), then $H_0$ can be rejected at the $(1 - \alpha)$ level. If $H_0$ is not rejected, then both $MS_{LOF}$ and $MS_{ME}$ are estimates of the measurement error and can be pooled to give a better estimate $\sigma^2 = (SS_{LOF} + SS_{ME})/(N - 2)$ with $N - 2$ degrees of freedom. The steps are given in table 8.4.

Table 8.4. How to validate the linearity of a calibration given $k$ calibration solutions each measured $n_i$ times

| Task | Equation |
|---|---|
| For each concentration ($i$), measured $n_i$ times: | |
|    Calculate the mean response | $\bar{y}_i$ |
|    Subtract mean response from each response at that concentration | $y_{ij} - \bar{y}_i$ |
| Square and sum for that concentration | $\sum_{j=1}^{n_i}\left(y_{ij} - \bar{y}_i\right)^2$ |
| Sum over all $k$ concentrations $= MS_{ME}$ | $\sum_{i=1}^{k}\sum_{j=1}^{n_i}\left(y_{ij} - \bar{y}_i\right)^2$ |
| For each concentration ($i$), measured $n_i$ times: | |
|    Subtract the mean response from the $y$ value estimated from the calibration | $\hat{y}_i - \bar{y}_i$ |
|    Square and multiply by the number of replicates | $n_i(\hat{y}_i - \bar{y}_i)^2$ |
| Sum over all $k$ concentrations $= MS_{LOF}$ | $\sum_{i=1}^{k} n_i\left(\hat{y}_i - \bar{y}_i\right)^2$ |
| Calculate the $F$ statistic | $F = \dfrac{MS_{LOF}}{MS_{ME}}$ |
| Calculate the probability ($P$) associated with $F$ for a one-sided distribution with $k - 2$ ($MS_{LOF}$) and $N - k$ ($MS_{ME}$) degrees of freedom. Reject the hypothesis of a linear relationship if $P < 0.05$ | =TDIST(F, k - 2, N - k) |

I work through the example presented in figure 8.7 here. Spreadsheet 8.1 shows the data with four measurements made at each of eight concentrations. The likely linear range is shown in the figure, and to illustrate the method, the range with one more point is tested first (to show it is not sufficiently linear) and then the more likely range is tested. Spreadsheet 8.2 shows the calculations for the points up to 50 nM.

Note some of the Excel shortcuts used. The command =SUMSQ(range) calculates the sum of the squares of the numbers in cells of the range. The command =SUMXMY2(range 1, range 2) calculates the sum of the squares of the differences between numbers in cells in the ranges specified, and =TREND(y-range, x-range, x-value, intercept) returns the calculated y-value from a linear fit of the y,x ranges given at the point x-value (which does not have to be one of the values in the range, or even in the range for that matter). The parameter intercept is 1 or TRUE if there is an intercept in the model, and 0 or FALSE if the line is forced through zero. The command =FDIST(f, df1, df2) returns the probability Pr(F>f) for the F distribution with df1 degrees of freedom in the numerator and df2 degrees of freedom in the denominator. The result is that the probability that the measurement error exceeds the lack of fit by chance is 0.014, so the hypothesis that the linear model holds can be rejected at the 98.6% level. The fit is graphed in figure 8.8, and it is clearly not a straight line.

All this is repeated in spreadsheet 8.3 and figure 8.9 for data up to 26.8 nM. Now the probability given by the F test is 0.55, well within acceptable limits.

### 8.6.3.3  Calibration Parameters (Sensitivity)

The sensitivity of a method (not to be confused with selectivity or limit of detection) is how much the indication of the measuring instrument increases

| | A | B | C | D | E | F |
|---|---|---|---|---|---|---|
| 34 | | | | | | |
| 35 | | | | | | |
| 36 | | | | | | |
| 37 | | | | | | |
| 38 | x /nM | | | y /microA | | ybar |
| 39 | 3.1 | 2.5 | 2.8 | 2.6 | 2.4 | 2.58 |
| 40 | 6.3 | 5.2 | 4.9 | 4.4 | 6 | 5.13 |
| 41 | 12.6 | 11 | 10 | 8.3 | 7.1 | 9.10 |
| 42 | 20.5 | 15 | 11 | 14 | 14 | 13.50 |
| 43 | 26.8 | 16 | 17 | 14 | 18 | 16.25 |
| 44 | 50.0 | 25 | 25 | 21 | 18 | 22.25 |
| 45 | 103.9 | 23 | 28 | 28 | 37 | 29.00 |
| 46 | 163.7 | 36 | 32 | 34 | 27 | 32.25 |
| 47 | | | | | | |

=AVERAGE(B39:E39)

**Spreadsheet 8.1.** Data for the analysis of copper by an electrochemical method. Each solution was measured four times. The data are plotted in figure 8.7.

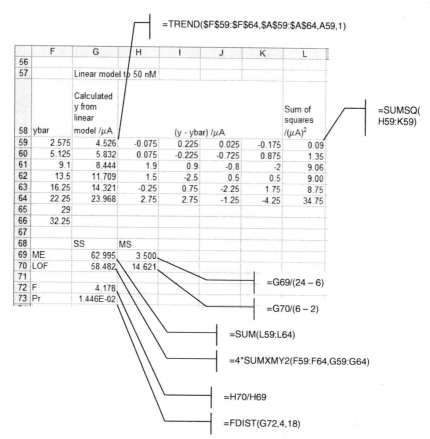

=TREND($F$59:$F$64,$A$59:$A$64,A59,1)

=SUMSQ(H59:K59)

| | F | G | H | I | J | K | L |
|---|---|---|---|---|---|---|---|
| 56 | | | | | | | |
| 57 | | Linear model to 50 nM | | | | | |
| 58 | ybar | Calculated y from linear model /μA | | | (y - ybar) /μA | | Sum of squares /(μA)² |
| 59 | 2.575 | 4.526 | -0.075 | 0.225 | 0.025 | -0.175 | 0.09 |
| 60 | 5.125 | 5.832 | 0.075 | -0.225 | -0.725 | 0.875 | 1.35 |
| 61 | 9.1 | 8.444 | 1.9 | 0.9 | -0.8 | -2 | 9.06 |
| 62 | 13.5 | 11.709 | 1.5 | -2.5 | 0.5 | 0.5 | 9.00 |
| 63 | 16.25 | 14.321 | -0.25 | 0.75 | -2.25 | 1.75 | 8.75 |
| 64 | 22.25 | 23.968 | 2.75 | 2.75 | -1.25 | -4.25 | 34.75 |
| 65 | 29 | | | | | | |
| 66 | 32.25 | | | | | | |
| 67 | | | | | | | |
| 68 | | SS | MS | | | | |
| 69 | ME | 62.995 | 3.500 | | | | |
| 70 | LOF | 58.482 | 14.621 | | | | |
| 71 | | | | | | | |
| 72 | F | 4.178 | | | | | |
| 73 | Pr | 1.446E-02 | | | | | |

=G69/(24 – 6)

=G70/(6 – 2)

=SUM(L59:L64)

=4*SUMXMY2(F59:F64,G59:G64)

=H70/H69

=FDIST(G72,4,18)

**Spreadsheet 8.2.** Calculations for test of linear range of data shown in spreadsheet 8.1. Range tested, 3.1–50.0 nM; $N = 24$ data points; number of concentrations, $k$, = 6. ME = measurement error, LOF = lack of fit, SS = sum of squares, MS = mean square.

with a given change in concentration. For a linear calibration, this is the slope of the calibration plot ($b$ in equation 8.1). In a more general calibration equation, for each term that includes a parameter times a function of $x$, the parameter is part of the sensitivity of the model. Only in linear calibration is the sensitivity constant across the range of the calibration. In other cases the sensitivity changes and must be judged accordingly. Figure 8.7 is an example of this situation. The fit to the nonlinear model is very good, but the slope of the line continually decreases and with it, the sensitivity. There comes a point when the calibration, although passing the tests described above, may be no longer be sufficiently sensitive to yield suitable results.

In linear calibrations (I use the term in its mathematical sense—linear in the parameters of $x$, allowing a quadratic calibration model $y = a + b_1x + b_2 x^2$

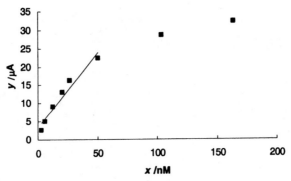

**Figure 8.8.** Linear fit of the of the data of spreadsheet 8.1 and figure 8.7 up to $x$ = 50.0 nM.

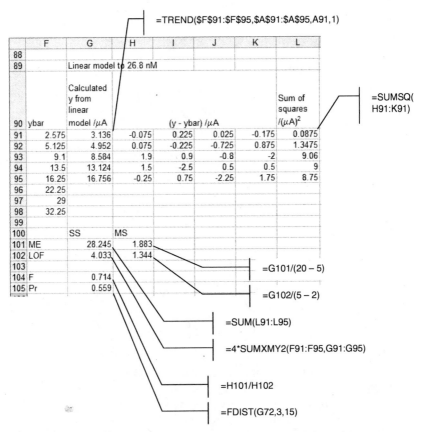

=TREND($F$91:$F$95,$A$91:$A$95,A91,1)

=SUMSQ(H91:K91)

| | F | G | H | I | J | K | L |
|---|---|---|---|---|---|---|---|
| 88 | | | | | | | |
| 89 | | Linear model to 26.8 nM | | | | | |
| 90 | ybar | Calculated y from linear model /μA | | (y - ybar) /μA | | | Sum of squares /(μA)² |
| 91 | 2.575 | 3.136 | -0.075 | 0.225 | 0.025 | -0.175 | 0.0875 |
| 92 | 5.125 | 4.952 | 0.075 | -0.225 | -0.725 | 0.875 | 1.3475 |
| 93 | 9.1 | 8.584 | 1.9 | 0.9 | -0.8 | -2 | 9.06 |
| 94 | 13.5 | 13.124 | 1.5 | -2.5 | 0.5 | 0.5 | 9 |
| 95 | 16.25 | 16.756 | -0.25 | 0.75 | -2.25 | 1.75 | 8.75 |
| 96 | 22.25 | | | | | | |
| 97 | 29 | | | | | | |
| 98 | 32.25 | | | | | | |
| 99 | | | | | | | |
| 100 | | SS | MS | | | | |
| 101 | ME | 28.245 | 1.883 | | | | |
| 102 | LOF | 4.033 | 1.344 | | | | |
| 103 | | | | | | | |
| 104 | F | 0.714 | | | | | |
| 105 | Pr | 0.559 | | | | | |

=G101/(20 – 5)

=G102/(5 – 2)

=SUM(L91:L95)

=4*SUMXMY2(F91:F95,G91:G95)

=H101/H102

=FDIST(G72,3,15)

**Spreadsheet 8.3.** Calculations for test of linear range of data shown in spreadsheet 8.1. Range tested, 3.1–26.8 nM; $N$ = 20 data points; number of concentrations, $k$, = 5. ME = measurement error, LOF = lack of fit, SS = sum of squares, MS = mean square.

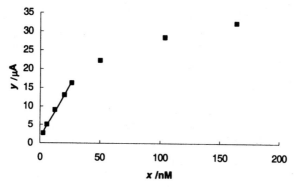

**Figure 8.9.** Linear fit of the of the data of spreadsheet 8.1 and figure 8.7 up to $x = 26.8$ nM.

to be classed as linear), it is possible to calculate the standard error on each parameter. In Excel this is easily done in LINEST (see chapter 2), or in the regression analysis section of the Data Analysis tools (Menu: Tools, Data Analysis, Regression).

### 8.6.3.4 Demonstration of Metrological Traceability

"Calibration" referred to here is the calibration, usually performed in the laboratory on a reasonably frequent basis, of the indication of the measuring instrument against concentrations of standards of the material to be measured. Use of certified reference materials (CRMs) to make up working calibration standards gives metrological traceability to the measurements that will be subsequently made by the calibrated system (assuming that the uncertainty of the values of the standards is properly incorporated into the measurement uncertainty). However, it must be remembered that other aspects of the system might also require calibration (e.g., glassware, balances, thermometers). Some of these are calibrated once in the factory, whereas others need yearly calibration. The performance of these calibrations and maintenance of proper records is part of the quality assurance system of the laboratory. See chapters 7 and 9 for further details.

### 8.6.4 Systematic Effects (Accuracy and Trueness)

The analyst conducting a method validation must assess any systematic effects that need to be corrected for or included in the measurement uncertainty of the results. The interplay between random and systematic error is complex and something of a moveable feast. The unique contribution of each analyst to the result of a chemical measurement is systematic, but in a large interlaboratory study with many analysts, all measuring aliquots of the same material, their individual biases will now be part of the reproducibility precision of the method (assuming there has been a genuinely random selection of analysts, whose biases are distributed normally). Figure 8.10 shows this interplay going from repeatability and run bias in a single measurement to reproducibility (inevitably greater) and method bias (usually lesser) in an interlaboratory study.

How these different effects are treated depends on the nature of the method validation. For the validation of a new method by an international body, including an extensive interlaboratory study, the method bias and reproducibility will be reported. For a single laboratory, laboratory bias and intermediate reproducibility will be determined.

Both within a single laboratory or in an interlaboratory study, the approach to assessing bias and recovery is similar. The method is assessed using either reference materials with known quantity values, or the analysis of a test material is compared with the results of analysis of the same material

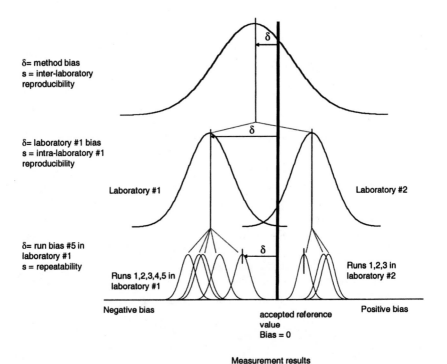

**Figure 8.10.** Bias and precision in chemical analysis. Contributions to bias when averaged by many repeated measurements become part of the variance of the measurement at the level (run, laboratory, method) being studied. (Adapted from O'Donnell and Hibbert 2005.)

by a reference method. Analysis of a single material yields a single estimate of bias, that, if significant, is usually subtracted from subsequent measurement results. This assumes linearity of the response, and in a complete study should be checked by analyzing a range of reference materials, which could lead to a bias correction with a constant and a proportional part.

When a CRM is used, at least 10 independently prepared portions of the material should be analyzed. Suppose the concentration of the standard solution made from the CRM is $c_{CRM}$ and the mean of the determinations is $c_{obs}$. If the uncertainty of the concentration of the CRM is $u_{CRM}$ (the uncertainty of the purity of the CRM combined with the uncertainty of making a solution to present to the instrument), then the uncertainty of the measurement of the standard solution is

$$u_{bias} = \sqrt{\frac{s_r^2}{10} + u_{CRM}^2}$$ (8.10)

where $s_r$ is the repeatability of the measurement. The difference $\delta = |c_{CRM} - c_{obs}|$ is tested against zero by a $t$ test:

$$t = \frac{|c_{CRM} - c_{obs}|}{u_{bias}}$$ (8.11)

at 9 degrees of freedom. If the null hypothesis is rejected and the analyst concludes that the bias is significant, then the bias ($\delta$) is reported and results are expected to be corrected for this value when the method is used in circumstances under which the bias estimate holds. For a reference material for which there is some confidence in the quantity value of interest, but that does not have a certificate and so does not qualify as a CRM, $u_{CRM}$ in equation 8.10 is unknown. Or a CRM might exist, but the matrix in which it will be used has not been studied, and there is concern about the recovery. In these cases it is still better to use whatever data are available to estimate bias. The uncertainty of the reference material $u_{RM}$ should not be written as zero, but a reasonable estimate should be inserted in equation 8.10 and the procedure documented and justified in the method validation documentation. There is a certain irony in testing for significant bias, for the greater the uncertainty in the measurement of bias, the less likely it is to find significant bias.

When a reference method is used to perform the bias estimate, a similar calculation is done, this time on the difference between the results of the method under validation and the reference method ($c_{ref}$). Now the uncertainty of the difference is that of two independent measurements, and assuming each was done 10 times with the repeatability of the reference method being $s_r(ref)$,

$$u_{bias} = \sqrt{\frac{s_r^2}{10} + \frac{s_r^2(ref)}{10}} \tag{8.12}$$

$$t = \frac{|c_{ref} - c_{obs}|}{u_{bias}} = \frac{|c_{ref} - c_{obs}| \times \sqrt{10}}{\sqrt{s_r^2 + s_r^2(ref)}} \tag{8.13}$$

with, again, 9 degrees of freedom for the 10 differences.

The term "recovery" is used to describe the fraction of a test material that can be extracted from a matrix. In analytical chemistry it also means the fraction of a test material in a matrix that can be quantified, even if it is not physically separated from the matrix. In a typical experiment, pure reference material is added to a matrix, mixing as completely as possible. The analysis of the material then gives the recovery (also known as surrogate recovery or marginal recovery). The procedure rests on the assumption that a spiked matrix has the same analytical properties as a matrix with native analyte. This might not always be true, and although observation of a recovery that is not unity strongly implies the presence of bias, a recovery of unity could hide bias in the analysis of native analyte (Thompson et al. 1999). The EURACHEM (1998) guide to validation recommends measurements of a matrix blank or an unfortified sample (i.e., one without any added analyte), followed by spiking with a known amount of analyte and reanalysis. This should be done six times starting from independent blanks or samples. If the measured concentration of analyte initially is $c_{blank}$ and after spiking $c_{fortified}$, and the concentration of the spike in the sample is known to be $c_{spike}$, then the recovery is calculated as

$$R\% = 100\% \times \frac{c_{fortified} - c_{blank}}{c_{spike}} \tag{8.14}$$

The usefulness of such a recovery rests on how well the spike is mixed with the matrix and how closely its properties in the matrix resemble those of the native analyte. Most guides recommend correction for recovery if it has been estimated in an appropriate way with known uncertainty (that can be included in estimation of measurement uncertainty). However, as recovery is different for different matrices, some international bodies (e.g., Codex Alimentarius for some food analyses) require results to be quoted as obtained without modification for recovery.

### 8.6.5 Measurement Uncertainty

Measurement uncertainty is the key to understanding modern approaches to quality assurance in analytical chemistry. A proper measurement uncertainty gives a range in which the value of the measurand is considered to exist with a high degree of probability (e.g., 95%), and this range includes contributions from all sources, including incomplete specification of the measurand, uncorrected systematic measurement effects and random measurement effects. Remember that measurement uncertainty is a property of a measurement result, not a method. Therefore, when a particular measurement uncertainty is revealed in a method validation study, before a laboratory decides to quote the value on test reports, it must ensure that the results it produces are consistent with this uncertainty. At the minimum, repeatability precision should be shown to conform to those values quoted and used in the uncertainty budget, and any aspect of the budget that differs should be highlighted and amended accordingly. The construction of an uncertainty budget and the nature of measurement uncertainty was described and discussed in chapter 6.

### 8.6.5.1    Type A Uncertainty
### Components (Precision)

Depending on the extent of the method validation, it may be possible to estimate repeatability standard deviation (at the very least); intermediate precision (done in the same laboratory, but over several days and possibly by different analysts and instruments); and/or reproducibility (full interlaboratory study). The most simple way to estimate repeatability and intermediate reproducibility is to analyze a test sample 10 times over a period of days, allowing variation in, for example, the analyst carrying out the measurement; the instrument used; the reagents used for calibration; and glassware, balances, and any other relevant influence factor. Of course, in a small laboratory with a single analyst and one GC, the scope for varying conditions may be limited, and so the precision obtained, although encouragingly small, will have a restricted scope and will need to be revisited if anything changes. Laboratories are encouraged to use all valid data to contribute to the estimation of measurement uncertainty. This should decrease the time needed to produce the uncertainty budget, and the uncertainty will have more relevance for measurements taken in a particular laboratory than estimated (Type B) components.

### 8.6.5.2    Type B Uncertainty Components

All sources of uncertainty that are not quantified by the standard deviation of repeated measurements fall in the category of Type B components. These were fully dealt with in chapter 6. For method validation, it is important to document the reasoning behind the use of Type B components because Type B components have the most subjective and arbitrary aspects. Which components are chosen and the rationale behind the inclusion or exclusion of components should documented. The value of the standard uncertainty and the distribution chosen (e.g., uniform, triangular, or normal) should be made available, as should the final method used to combine all sources.

### 8.6.5.3 Combined and
### Expanded Uncertainty

The major components of uncertainty are combined according to the rules of propagation of uncertainty, often with the assumption of independence of effects, to give the combined uncertainty. If the measurement uncertainty is to be quoted as a confidence interval, for example, a 95% confidence interval, an appropriate coverage factor is chosen by which to multiply the combined uncertainty and thus yield the expanded uncertainty. The coverage factor should be justified, and any assumptions about degrees of freedom stated.

Measurement uncertainty is a property of a measurement result, and so an estimate made during method validation is only valid if, when real measurements are made, the analyst can assert that the conditions of his or her test exactly follow those under which the validation was performed.

### 8.6.6  Ruggedness and Robustness

Method validation covers a number of aspects of an analytical method that have already been evaluated in the course of development and use. The values of the calibration parameters must be known to use the method to analyze a particular sample, and any serious deviations from the measurement model should have been discovered. In addition, however, every method should undergo a robustness study as the practicality of the method may ultimately depend on how rugged it is.

A method is classed as rugged if its results remain sufficiently unaffected as designated environmental and operational conditions change. Exactly which conditions are singled out for attention in a robustness study is a matter for the analyst's judgment, perhaps in consultation with the client. Also, what constitutes "sufficiently unaffected" must be defined before the method validation experiments are done. A robustness study addresses two areas of concern: the need for a method to be able to yield acceptable results under the normal variation of conditions expected during routine operation, and the portability of the method between laboratories with changes in instrumentation, people, and chemical reagents. A method being developed in house in a pharmaceutical company for process control might focus on the first area, whereas a method to be published as an international standard will need to address both areas.

In a robustness study the parameters to be investigated are systematically changed and the effects on the result of the analysis are measured. The steps in such a study are shown in table 8.5.

### 8.6.6.1  Parameters to Be Studied

Table 8.5 gives some typical parameters that might be studied for ruggedness. Judgment is required because, as the number of parameters studies

Table 8.5. Steps in a robustness study

| Action | Examples |
| --- | --- |
| Identify parameters to be studied | Temperature, reagent source, instrument, column age, column type, pH, mobile phase, detector type |
| Decide on statistical model | Plackett-Burman design; fractional factorial design |
| Decide on range of each parameter | $T$ = 10–30 °C, pH = 5–8; new column– column used 1 month; methanol– water ratio in mobile phase varied by ±2 % |
| Perform randomized experiments according to statistical model | Run 1: $T$ = 25 °C, pH = 7, new column, MeOH/water = 30% . . . |
| Calculate effects of each factor and test for statistical significance | Effect ($T$) = +2.3%, effect (pH) = -1.0% |

increases, the number of experiments that must be performed also rises, and with some statistical models this may be expressed as a power law (e.g., $2^k$ for a full two-level factorial design; see chapter 3). There is no point in investigating a factor that is well known to have no effect, nor one that is important but is well controlled as part of the experiment. During method development the adequacy of, for example, temperature control in a GC oven, should have been thoroughly checked. However, if during the course of a measurement the laboratory temperature varies between 15° and 28°C, then it might be prudent to investigate whether this variation has any effect on the results of the experiment. The robustness of the method with regard to the precision of weighing or the volume of an added reagent can be investigated by relaxing the requirements of the method. For example, if 5 g of a reagent must be added, does this mean 5.0000 g or somewhere between 4.5 g and 5.5 g?

Choice of the levels of the parameters is important, and making poor choices can render a study useless. Most statistical models used for robustness studies use only two levels (usually written in designs as contrast coefficients + and –; see chapter 3). These can be chosen as the smallest and greatest values of the parameter expected during normal use of the method, or the nominal value specified in the method (conventionally ascribed to –) and a changed value (+). It is important not to designate values that are outside any sensible range. The results will not be interpretable, and a real effect can be masked. In the example in figure 8.11 (which is a hypothetical pH response from chapter 3), by choosing pH values outside a valid range (a,b), the result of the analysis for each level is unnaturally low, and the difference between the results is not statistically different from zero. An analyst who is not alert to the problem (and if this analyst chose the levels in the first place, he or she would no doubt remain in blissful ignorance)

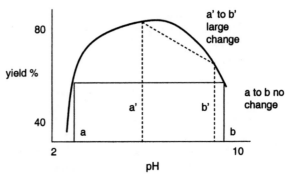

**Figure 8.11.** A hypothetical effect of changing pH
on the result of an analysis, as the subject of a
robustness study. (a) Inappropriate choice of
levels leading to an apparent lack of effect. (b) A
more sensible choice with a significant main
effect being calculated.

would accept the result of the statistics that the effect was near zero. A more
realistic pair of pH values (a′,b′), where the first is the nominal value at which
the experiment is usually performed, leads to a useful answer.

### 8.6.6.2  Statistical Model

In a desire to minimize the work required to establish the ruggedness of the
method, statistical corners will be cut in establishing the effect of changing
the parameters of interest. The Plackett-Burman experimental design (chapter 3) is a commonly used method that shares many of the basic assumptions of highly fractionated designs. It has the advantage that only one more
experiment must be done than the number of factors being investigated,
which makes it very efficient. The outcome of the experimental design is
estimates of the main effects of the factors; that is, by how much does the
response variable change as the factor is changed from its low (normal) value
to its high value? If this is shown by statistics to be insignificant (i.e., not
different from zero within the variability of the measurements), then it can
be concluded that the method is robust to that factor. Full details of how to
do a Plackett-Burman design were given in chapter 3.

## 8.7  Documentation

Method validation is documented for two reasons. The first is to allow a user
of the method to understand the scope of validation and how to perform
experiments within the scope of the validation. As soon as the method is
used outside the scope of its validation, the onus is on the analyst to com-

plete and document any necessary additional validation. Again, experience is what allows an analyst to know that the scope has been widened but not widened enough to invalidate a previous method validation. As long as this decision has been documented, requirements have been satisfied.

The second reason for documenting method validation is that the method is not validated until it it is documented. No number of experiments is of any use if they are not documented for future users and clients to consult. The definitions of method validation, with which this chapter started, mostly refer to documentary evidence. The extent of the documentation, both in describing the method and its scope, and the method validation experiments, is a matter of judgment, but a general rule is to include more, rather than less, information. What the analyst who has painstakingly developed and validated the method may take for granted could be a crucial piece of information for a subsequent user. The ISO 78-2 guide for documenting chemical analysis standards may be of use, even if the method is being developed for more humble and restricted uses (ISO 1999). As the documentation will become part of the laboratory's quality assurance system, proper documentary traceability (as opposed to metrological traceability of a result) must be maintained, with revision versions and appropriate authorizations for updates.

The Laboratory of the Government Chemist has published software to guide a would-be validator through the required steps (LGC 2003). The program then can produce a report with all relevant information, formatted in a style approved by the LGC.

## References

Buck, R P and Lindner, E (1994), Recommendations for nomenclature of ion-selective electrodes. *Pure and Applied Chemistry*, 66, 2527–36.

Burgess, C (2000), *Valid analytical methods and procedures: a best practice approach to method selection, development and evaluation* (Cambridge: Royal Society of Chemistry).

Christensen, J M, et al. (1995), Method validation: an essential tool in total quality management, in M Parkany (ed.), *Quality assurance and TQM for analytical laboratories* (Cambridge: Royal Society of Chemistry), 46–54.

Currie, L A (1995), Nomenclature in evaluation of analytical methods including detection and quantification capabilities. *Pure and Applied Chemistry*, 67, 1699–723.

DWI (Drinking Water Inspectorate) (1993), Further guidance on analytical systems (Department of the Environment London), Available: http://www.dwi.gov.uk/regs/infolett/1993/info893.htm.

Ellison, S L R, Gregory, S and Hardcastle, W A (1998), Quantifying uncertainty in qualitative analysis, *The Analyst*, 123, 1155–61.

EURACHEM (1998), Eurachem guide: the fitness for purpose of analytical methods (Teddington, UK: Laboratory of the Government Chemist).

Fajgelj, A and Ambrus, A (eds.) (2000), *Principles and practices of method validation* (Cambridge: Royal Society of Chemistry).

Green, J M (1996), A practical guide to analytical method validation. *Analytical Chemistry*, 305–9A.

Hibbert, D B (2005), Method validation, in P J Worsfold, A Townshend, and C F Poole (eds.), *Encyclopedia of analytical science*, 2nd ed., vol. 7 (Oxford: Elsevier), 469–77.

Horwitz, W (1995), Revised protocol for the design, conduct and interpretation of method performance studies. *Pure and Applied Chemistry*, 67, 331–43.

ICH (1995), Text on validation of analytical procedures, Q2A (Geneva: International Conference on Harmonization of Technical Requirements for Registration of Pharmaceuticals for Human Use).

ICH (1996), Validation of analytical procedures: methodology, Q2B (Geneva: International Conference on Harmonization of Technical Requirements for Registration of Pharmaceuticals for Human Use).

ISO (1994a), Quality management and quality assurance—Vocabulary, 8402 (Geneva: International Organization for Standardization).

ISO (1994b), Accuracy (trueness and precision) of measurement methods and results—Part 2: Basic method for the determination of repeatability and reproducibility of a standard measurement method 5725-2 (Geneva: International Organization for Standardization).

ISO (1999), Chemistry—Layouts for standards—Part 2: Methods of chemical analysis, 78–2 (Geneva: International Organization for Standardization).

ISO (2005), Quality management systems—Fundamentals and vocabulary, 9000 (Geneva: International Organization for Standardization).

ISO/IEC(1997), Proficiency testing by interlaboratory comparisons: Part 1— Development and operation of proficiency testing schemes, Guide 43-1 (Geneva: International Organization for Standardization).

ISO/IEC (2005), General requirements for the competence of calibration and testing laboratories, 17025 2nd ed. (Geneva: International Organization for Standardization).

LGC (2003), mVal—Software for analytical method validation (Teddington, UK: Laboratory of the Government Chemist).

Long, G L and Winefordner, J D (1983), Limit of detection—A closer look at the IUPAC definition. *Analytical Chemistry*, 55, 712A–24A.

O'Donnell, G E and Hibbert, D B (2005), Treatment of bias in estimating measurement uncertainty, *The Analyst,* 130, 721–29.

Ong, R C Y and Marriott, P J (2002), A review of basic concepts in comprehensive two-dimensional gas chromatography, *Journal of Chromatographic Science*, 40 (5), 276–91.

Thompson, M, Ellison, S, and Wood, R (2002), Harmonized guidelines for single laboratory validation of methods of analysis. *Pure and Applied Chemistry*, 74, 835–55.

Thompson, M, et al. (1999), Harmonized guidelines for the use of recovery information in analytical measurement, *Pure and Applied Chemistry*, 71 (2), 337–48.

USP (1999), Validation of compendial methods, in *The United States Pharmacopeia 24—National Formulary 19, General Chapter 1225* (Rockville, MD: United States Pharmacopeial Convention, Inc.).

Wood, R (1999), How to validate analytical methods, *Trends in Analytical Chemistry*, 18, 624–32.

Zhao, M, Hibbert, D B, and Gooding, J J (2003), Solution to the problem of interferences in electrochemical sensors using the fill-and-flow channel biosensor. *Analytical Chemistry* 75 (3), 593–600.

Zhao, M, et al. (2004), A portable fill-and-flow channel biosensor with an electrode to predict the effect of interferences. *Electroanalysis*, 16 (15), 1221–26.

# 9

## Accreditation

### 9.1 Introduction

Accreditation is the procedure by which the competence of a laboratory to perform a specified range of tests or measurements is assessed against a national or international standard. The accreditation covers the kinds of materials tested or measured, the procedures or methods used, the equipment and personnel used in those procedures, and all relevant systems that the laboratory has in place. Once accredited, the laboratory is entitled to endorse test results with their accreditation status which, if it has any validity, is an imprimatur of some degree of quality and gives the client added confidence in the results. Accreditation therefore benefits the laboratory, by allowing the laboratory to demonstrate competence in particular tests, and the client, by providing a choice of accredited laboratories that are deemed competent.

### 9.1.2 The Worldwide Accreditation System

Accreditation is part of conformity assessment in international trade. Conformity assessment leads to the acceptance of the goods of one country by another, with confidence borne of mutual recognition of manufacturing and testing procedures. Figure 9.1 shows the relation between accreditation and the goal of conformity in trade. For accreditation to be a cornerstones of

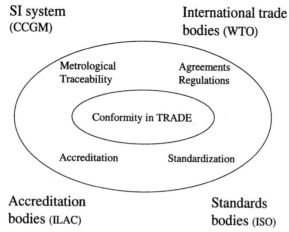

SI system
(CCGM)

International trade
bodies (WTO)

Metrological
Traceability

Agreements
Regulations

Conformity in TRADE

Accreditation

Standardization

Accreditation
bodies (ILAC)

Standards
bodies (ISO)

**Figure 9.1.** Relationships between the international conformity in trade system and quality systems including accreditation. (Adapted from a document of European Accreditation, accessed from www.european-accreditation.org, October 2004.)

conformity in trade, each laboratory that is assessed, in whatever country, must be judged against the same standard (e.g., ISO/IEC 17025), and the assessment process must be essentially the same from one country to another. The standards are indeed international, through the International Organization for Standardization (ISO), and the accreditation bodies themselves are scrutinized under the auspices of the International Laboratory Accreditation Co-operation (ILAC), being accredited to the ISO/IEC Standard 17011 (ISO/IEC 2004a).

Full membership in ILAC is open to recognized bodies that operate accreditation schemes for testing laboratories, calibration laboratories, and inspection bodies that have been accepted as signatories to the ILAC Mutual Recognition Arrangement. They must maintain conformance with appropriate international standards such as ISO/IEC 17011 and ILAC guidance documents, and the must ensure that all their accredited laboratories comply with ISO/IEC 17025 and related ILAC guidance documents. Table 9.1 lists the full members and signatories of the ILAC Mutual Recognition Arrangement. The National Association of Testing Authorities (NATA) of Australia has the distinction of being the first accreditation body in the world (founded in 1947), and has long been in the vanguard of the approach to quality through accreditation.

There are also regional groups covering Europe and the Middle East (European Accreditation for Cooperation – EA), the Americas (Interamerican Accreditation Cooperation – IAAC), the Asia Pacific region including India

Table 9.1. Accreditation bodies who are full members of International Laboratory Accreditation Co-operation (ILAC) and signatories to the MLA

| Country/ region | Accreditation body | Acronym and web address |
|---|---|---|
| Argentina | Organismo Argentino de Acreditacion | OAA http://www.oaa.org.ar |
| Australia | National Association of Testing Authorities | NATA http://www.nata.asn.au/ |
| Austria | Bundesministerium fur Wirtschaft und Arbeit | BMWA http://www.bmwa.gv.at |
| Belgium | Belgian Accreditation Structure | BELAC (BELTEST/ BKO) http://BELAC.fgov.be |
| Brazil | General Coordination for Accreditation | CGCRE/INMETRO http://www.inmetro.gov .br |
| Canada | Canadian Association for Environmental Analytical Laboratories | CAEAL http://www.caeal.ca |
| Canada | Standards Council of Canada | SCC http://www.scc.ca |
| Peoples Republic of China | China National Accreditation Board for Laboratories | CNAL http://www.cnal.org.cn |
| Cuba | National Accreditation Body of Republica de Cuba | ONARC http://www.onarc .cubaindustria.cu |
| Czech Republic | Czech Accreditation Institute | CAI http://www.cai.cz |
| Denmark | Danish Accreditation | DANAK http://www.danak.org |
| Egypt | National Laboratories Accreditation Bureau | NLAB http://www.nlab.org |
| Finland | Finnish Accreditation Service | FINAS http://www.finas.fi |
| France | Comité Francais d'Accreditation | COFRAC http://www.cofrac.fr/ |
| Germany | German Accreditation Body Chemistry | DACH http://dach-gmbh.de/ |
| Germany | German Accreditation System for Testing | DAP http://www.dap.de |
| Germany | German Accreditation Body Technology | DATech http://www.datech.de |
| Germany | Deutscher Kalibrierdienst | DKD http://www.dkd.info |
| Greece | Hellenic Accreditation System S.A. | ESYD http://www.esyd.gr |
| Hong Kong | Hong Kong Accreditation Service | HKAS http://www.itc.gov.hk/ hkas |

Table 9.1. *(continued)*

| Country/ region | Accreditation body | Acronym and web address |
|---|---|---|
| India | National Accreditation Board for Testing & Calibration Laboratories | NABL http://www.nabl-india .org |
| Ireland | Irish National Accreditation Board | NAB http://www.inab.ie |
| Israel | Israel Laboratory Accreditation Authority | ISRAC http://www.israc.gov.il |
| Italy | Sistema Nazionale per l'Accreditamento di Laboratori | SINAL http://www.sinal.it |
| Italy | Servizio di Taratura in Italia | SIT http:// www.sit-italia.it/ |
| Japan | International Accreditation Japan | IA Japan http://www.nite.go.jp/ asse/iajapan/en/ |
| Japan | The Japan Accreditation Board for Conformity Assessment | JAB http://www.jab.or.jp |
| Korea, Republic of | Korea Laboratory Accreditation Scheme | KOLAS http://kolas.ats.go.kr |
| Mexico | Entidad Mexicana de Acreditación | EMA http://www.ema.org.mx |
| Netherlands, The | Raad voor Accreditatie | RvA http://www.rva.nl |
| New Zealand | International Accreditation New Zealand | IANZ http://www.ianz.govt.nz |
| Norway | Norsk Akkreditering | NA http://www.akkreditert.no |
| Phillipines | Bureau of Product Standards Laboratory Accreditation Scheme | BPSLAS http://www.dti.gov.ph |
| Poland | Polish Centre for Accreditation | PCA http://www.pca.gov.pl |
| Romania | Romanian Accreditation Association | RENAR http://www.renar.ro |
| Singapore | Singapore Accreditation Council | SAC http://www. sac-accreditation.org.sg |
| Slovakia | Slovak National Accreditation Service | SNAS http://www.snas.sk |
| Slovenia | Slovenian Accreditation | SA http://www.gov.si/sa |
| South Africa | South African National Accreditation System | SANAS http://www.sanas.co.za |

*(continued)*

Table 9.1. (*continued*)

| Country/ region | Accreditation body | Acronym and web address |
| --- | --- | --- |
| Spain | Entidad Nacional de Acreditacion | ENAC http://www.enac.es |
| Sweden | Swedish Board for Accreditation and Conformity Assessment | SWEDAC http://www.swedac.se |
| Switzerland | Swiss Accreditation Service | SAS http://www.sas.ch |
| Taipei, Taiwan | Taiwan Accreditation Foundation | TAF http://www.taftw.org.tw |
| Thailand | The Bureau of Laboratory Quality Standards | BLQS-DMSc http://www.dmsc.moph .go.th |
| Thailand | Thai Laboratory Accreditation Scheme/Thai Industrial Standards Institute | TLAS/TISI http://www.tisi.go.th |
| United Kingdom | UK Accreditation Service | UKAS http://www.ukas.com/ |
| United States | American Association for Laboratory Accreditation | A2LA http://www.a2la.org/ |
| United States | International Accreditation Service, Inc | IAS http://www.iasonline.org |
| United States | National Voluntary Laboratory Accreditation Program | NVLAP http://www.nist.gov/ nvlap |
| Vietnam | Vietnam Laboratory Accreditation Scheme | VILAS http://www.boa.gov.vn |

Taken from http://www.ilac.org/ (January 2006).

(Asia Pacific Laboratory Accreditation Cooperation—APLAC) and Africa (Southern African Development Community in Accreditation—SADAC). Through these cooperations the importance of harmonization and conformity has been recognized. If countries adopt similar approaches to accreditation, and their accreditation organizations issue certificates with a harmonized information, it is much easier for clients to assess the validity of test results relating to cross-border trade, forensics, and health.

### 9.1.3   Peer-led Accreditation

Peer assessment is the cornerstone of many accreditation systems. The accreditation body maintains a staff of administrators and experts, but the majority of the work of assessments, including visits to the laboratories seeking accreditation, is performed by laboratory scientists, often volunteers, who have undergone assessment training and who have a high level of technical competence. An assessment team is usually led by a permanent member of the accreditation organization staff to provide appropriate guidance and to help maintain equality of treatment across the activities of the organization. The remaining members are peers, chosen to include the desired spread of expertise. The team spends a couple of days at the laboratory, observing the work of the laboratory, scrutinizing documentation and procedures, and interviewing staff. The outcome of the visit is a report detailing the conformity of the laboratory's practices to the standard and indicating where improvements need to be made before the laboratory is recommended for accreditation. The recommendation, documented with the results of the visit, is assessed by a committee of the organization. The laboratory is then allowed to display the logo of the accreditation body on its certificates of analysis and is entered in a register of accredited laboratories for the tests or measurements that fall within the scope of the accreditation.

### 9.1.4    What a Standard Looks Like

Standards tend to have a common look and feel that can be confusing when first encountered. In fact, there is a standard on how to write standards, with which good standards comply (ISO 1999). Most start with three sections (called clauses), scope, normative references, and terms and definitions. The scope of a standard details what is covered by the standard, and just as important, what is outside the standard. Normative references are to other standards that are referenced by the present standard and which therefore must be adhered to as part of compliance with the present standard. Guidance is usually given to the parts of, or extent to which, these normative references must be followed. "Terms and definitions" are what they imply, but because much of the terminology of metrology is already well standardized, this clause is often just a reference to the international vocabulary (VIM; ISO 1993) or to ISO/IEC Guide 2 (ISO/IEC 2004b), with one or two additional specialist terms.

## 9.2  Accreditation to ISO/IEC 17025:2005

The ISO/IEC 17025 standard (ISO/IEC 2005) has the title "General requirements for the competence of testing and calibration laboratories" and is the main standard to which analytical chemical laboratories are accredited. The word "calibration" in the title arises from the use of the standard to accredit bodies that calibrate instruments such as balances, electrical equipment, and utility meters. It must also be stressed that the standard is not written for analytical chemists, but for any measurement scientists. Therefore, the terminology tends to be general, and the emphasis is sometimes not clear for the chemist. However, the standard is wide ranging and covers a whole community of measurement scientists.

The standard was first published in 1978 as ISO Guide 25 (ISO 1978), and later as the standard ISO/IEC 25 (ISO/IEC 1990). It was revised in 1994, and then underwent a major revision (van der Leemput 1997) and number change in 1999 when it became ISO/IEC 17025 (ISO/IEC 1999). Further tweaking to ensure alignment with ISO 9001:2000, and other technical amendments were made in 2005 (ISO/IEC 2005). ILAC has published a guide to accreditation under ISO/IEC 17025 (ILAC 2001), as have many national accreditation bodies.

In revising ISO Guide 25 in 1999, the management requirements were aligned with those in the ISO 9000 series, and conformed with ISO 9001:1994 (ISO 1994a) and 9002:1994 (ISO 1994b). The 2005 revision brought the standard into line with the single ISO 9001:2000 (ISO 2000). Appendix A in ISO/IEC 17025 gives nominal cross-references between the sections of the standards. A laboratory that is accredited to ISO/IEC17025 therefore also meets the requirements of ISO 9001 when it designs and develops new or nonstandard methods or when it only uses standard methods. ILAC allows a statement to this effect to be included in the accreditation certificate issued by a national accreditation body.

These notes and comments on different sections of the standard should be read in conjunction with the standard. The standard is available from ISO or from a national standards body in countries that have adopted the standard as part of a national program. (For example, in Australia ISO/IEC17025 is known as AS17025 and has been adopted without change.) ISO/IEC 17025 is a full international standard, but for many years its predecessor, Guide 25 was used, and many analysts and texts still refer to it as a "guide." A detailed discussion of the sections of the standard follows.

### 9.2.1  Scope

The scope of the 17025 standard states that it "specifies the general requirements for the competence to carry out tests and/or calibrations, including sampling. It covers testing and calibration performed using standard methods, non-standard methods, and laboratory-developed methods" (ISO/IEC 2005, Preface p ii).

Of interest is the inclusion of sampling in the 1999 revision of the standard. In the original, the laboratory was expected to analyze material as received. The laboratory was discouraged from talking to clients about their requirements or from offering an opinion once results had been obtained. Sampling is covered, for specific types of material, in a number of standards, but it is now recognized that if sampling must be done (i.e., the whole of the system under investigation cannot be tested, and therefore a representative portion must be analyzed), then the analyst may have to take the representative sample. If the material presented for analysis is not representative, then no matter how good the measurement is, it is likely to be of little use to the client. Therefore, if sampling is included in the scope of the analysis, it must also come under the scrutiny of assessment for accreditation.

The second part of the scope refers to the kind of method used and makes clear that whether the method is an internationally recognized standard or simply and in-house procedure, it falls within the scope of ISO/IEC 17025. In specifying exactly what a laboratory is accredited for, the accreditation body must be careful to correctly detail the methods. So if sulfur is analyzed by ASTM D5453, the laboratory must have demonstrated its capability to follow the method as laid specified in the American Society for Testing and Materials standard, or any changes made by the laboratory must be documented and validated.

### 9.2.2   Section 4:
### Management Requirements

Table 9.2 gives the subsections of section 4 of the ISO/IEC 17025 standard that deal with requirements for the management of the laboratory and the quality system that must be in place.

#### 9.2.2.1   Subsection 4.1: Organization

The first requirement is that the organization exists and is legally accountable. This applies to small enterprises that must be operating as a legal busi-

Table 9.2. Sections of ISO/IEC 17025:2005 dealing with management requirements

| Subsection | Subject |
|---|---|
| 4.1 | Organization |
| 4.2 | Management system |
| 4.3 | Document control |
| 4.4 | Review of requests, tenders and contracts |
| 4.5 | Subcontracting of tests and calibrations |
| 4.6 | Purchasing services and supplies |
| 4.7 | Service to the customer |
| 4.8 | Complaints |
| 4.9 | Control of non-conforming testing and/or calibration work |
| 4.10 | Improvement |
| 4.11 | Corrective action |
| 4.12 | Preventive action |
| 4.13 | Control of records |
| 4.14 | Internal audits |
| 4.15 | Management reviews |

ness, but also subsidiaries of international companies must have identifiable owners and structure. When accreditation is given, exactly what is being accredited must be known. Usually a laboratory or section operating in a particular location and with particular staff will be accredited. If an organization has separate laboratories in different cities operating with their own staff and local management, it is probable that each would need to have separate accreditation, even for the same analysis.

Another aspect of organizational management is the need for proper business procedures. The laboratory should not undertake work that results in a conflict of interest, and generally should not bring the accreditation system into disrepute by engaging in activities that would diminish confidence in the laboratory.

There is emphasis on the organizational structure, and the 2005 standard requires that the organization "shall ensure that its personnel are aware of the relevance and importance of their activities." (ISO/IEC 2005, section 4.1.5 k)

### 9.2.2.2 Subsection 4.2:
### Management System

A change in emphasis between 1999 and 2005 is the understanding that a quality system only exists in the context of the organization's management system. Thus the standard requires that the laboratory establish and maintain an appropriate management system. What constitutes appropriate should be obvious to the laboratory and must always be consistent with the ability of the laboratory to assure its quality. The commitment and responsibility of top management is stressed in this subsection. In particular, the top management must provide evidence of commitment to developing and implementing the system and must communicate to the organization the importance of meeting customer needs and statutory and regulatory requirements. The quality manual is the defining document and must start with a clear statement of the quality system policy and objectives. Limited discussion of technical aspects of quality control is found in subsection 5.9 of the standard.

### 9.2.2.3 Subsection 4.3: Document Control

Laboratories should have a procedure for control of documents, including manuals, software, drawings, specifications, and instructions. An authorized person must be responsible for approving and reviewing documents, and there must be a provision for removing out-of-date documents. To accomplish this, all documents must be uniquely identified with date of issue, page numbering and total pages, marked end page, and issuing authority. When documents are updated or changed, they should be subjected to the same review and approval procedures.

### 9.2.2.4   Subsection 4.4: Review of Requests, Tenders, and Contracts

A laboratory must have procedures for the review of requests, tenders, and contracts. Requirements must be defined against which tenders can be judged, and once in place they should be reviewed periodically. Before agreeing to any contract, the laboratory should be satisfied that the work can be done in accordance with the contract. Once a contract is signed, if there are significant changes, these must be documented and agreed-upon with the customer. Within the organization, staff involved in carrying out work must be informed of any changes in the contract.

### 9.2.2.5   Subsection 4.5: Subcontracting Tests

A laboratory may outsource aspects of its work occasionally because of unforeseen reasons (too great a workload, equipment problems, need for outside expertise) or on a regular basis. Where work is subcontracted, a competent laboratory (i.e., one that complies with the standard), must be used. The customer must be notified in writing about the use of subcontractors, and a register of all subcontractors, including calibration services, must be maintained.

### 9.2.2.6   Subsection 4.6: Purchasing Services and Supplies

The laboratory must have policy and procedures for all purchasing requirements, whether laboratory consumables, equipment, or calibration services. Records must be kept and any checks to ensure the products are fit for purpose documented. Records of purchases should be sufficiently detailed to identify what was purchased, when it was purchased, and any other relevant information, and the management system standard under which they were made should be identified.

### 9.2.2.7   Subsection 4.7: Service to the Customer

The laboratory must allow the customer reasonable access to the work, allow the customer to be present during testing, and give the customer access to any data and quality records in relation to the work. This must be done while ensuring confidentiality to other customers. It is now a requirement that the laboratory seek feedback from its customer and use that information to improve its service. Instruments for obtaining feedback might include customer satisfaction surveys.

### 9.2.2.8  Subsection 4.8: Complaints

The laboratory must maintain records of complaints, together with investigations and corrective actions.

### 9.2.2.9  Subsection 4.9: Control of Nonconforming Testing or Calibration Work

Nonconformance arises when any aspect of the work of a laboratory does not conform to its procedures, to the standard, or the agreed-upon requirements of the customer. These are detected by customer complaints, internal quality control measures, management audits and reviews, and by observations of staff. They must be dealt with under the appropriate part of the management system and fully investigated and reported.

### 9.2.2.10  Subsection 4.10: Improvement

New in the 2005 standard is a sentence that exhorts laboratories to continually improve the effectiveness of their management system.

### 9.2.2.11  Subsection 4.11: Corrective Action

Subsection 4.11 of the standard gives details the approach for identifying the root causes of nonconformance and explains how to select and implement actions to correct the problem.

### 9.2.2.12  Subsection 4.12: Preventive Action

Recognizing that heading off a potential problem is usually better (cheaper and less embarrassing) than coping with a nonconformance, the standard requires the laboratory to engage in preventive actions. This is defined as a proactive process to identify opportunities for improvement. The use of trend and risk analysis of proficiency testing data (see chapter 5) is mentioned as a possible preventive action.

### 9.2.2.13  Subsection 4.13: Control of Records

All activities carried out under the management system must be recorded and documented to facilitate retrieval as part of management, external review, or an audit. These documents must be identified and indexed and must be kept securely for a defined period. Electronic records must be secure, available only to properly identified personnel, and appropriately backed up. Technical records of original observations, derived data, calibrations, and calculations must be kept in a form so as to record the information, identify who made the observations, and show when the work was done. Any changes to these records must be done in such a way as to allow identification of the original record, the change, the person who made the change, and reason for the change.

The laboratory notebook with numbered pages, dated and signed by the member of staff and countersigned by a supervisor, is the traditional way of maintaining original observations. However, more and more instruments are computer controlled, and their hardcopy outputs are a brief synopsis of the result, with original data being held within the instrument's computer or captured by a laboratory information management system. The operation of any computer-controlled equipment will be covered by the validation of the method using the equipment, and this should include data handling. The management system must identify when protocols for storing and retrieving computer data are required.

### 9.2.2.14    Subsection 4.14: Internal Audits

An internal audit is designed to check compliance with the laboratory's management system and ISO/IEC 17025, to determine the effectiveness of the activity or process, and to identify improvements. An audit of the main elements of the management system should be carried out at least annually and should include some of the analytical methods. The schedule of audits is part of the management system. Trained and qualified auditors should be used, and, where possible, they should be independent of the activity being audited. Personnel from the organization can be used as long as they are not directly involved with the work being audited. The internal audit is one of the main drivers for continuous improvement and preventive action because it identifies areas for attention.

Many countries use volunteer peer reviewers for accreditation inspections. Although the commitment required can be seen as a cost to the reviewers' organizations, a benefit is the training they receive, which can be put to effective use in internal audits.

### 9.2.2.15    Subsection 4.15:
### Management Reviews

The management system is owned by and is the responsibility of the top management. It is incumbent on the management to review the overall effectiveness and currency of the system and the range of services offered by the laboratory. This should be done annually, and should include input from internal audits, clients, staff, and quality control records. The review should also consider supplier performance, wider management issues, and any changes that have affected the laboratory. The review should be used to identify areas for improvement and to set goals and action plans for the coming year.

### 9.2.3  Section 5: Technical Requirements

Table 9.3 gives the subsections of section 5 of the ISO/IEC 17025 standard, which details technical requirements. References to testing and measurement and a focus on how tests should be performed and reported make this part of the standard unique and of value to analytical chemists.

#### 9.2.3.1  Subsection 5.1: General

The standard recognizes the factors that determine the correctness and reliability of test results: human factors, accommodation and environment, methods, equipment, sampling, and the handling of test items. In this list, "measurement traceability" is mentioned, but in fact metrological traceability, with measurement uncertainty and method validation, are really subsumed in "methods." (subsection 5.4). The effect of each of these factors on measurement uncertainty will differ considerably among kinds of tests.

#### 9.2.3.2  Subsection 5.2: Personnel

It is heartening that the writers of the standard understand that chemical analysis is a human activity, and that the competence of staff is at the core of a successful laboratory. In chapter 5 I described some international interlaboratory studies showing there was little correlation between correctness of results and any identifiable factor (including accreditation), leading to the conclusion that the most important influence on the result is the ability of the analyst performing the test. The standard requires appropriately quali-

Table 9.3. Sections of ISO/IEC 17025:2005 dealing with technical requirements

| Subsection | Subject |
|---|---|
| 5.1 | General |
| 5.2 | Personnel |
| 5.3 | Accommodation and environmental conditions |
| 5.4 | Test and calibration methods and method validation |
| 5.5 | Equipment |
| 5.6 | Measurement traceability |
| 5.7 | Sampling |
| 5.8 | Handling of test and calibration items |
| 5.9 | Assuring the quality of test and calibration results |
| 5.10 | Reporting the results |

fied staff. The qualification may be education and training, experience, or demonstrated skills. Within a laboratory there might be a range of skills, but the manager or person who is the approved signatory will usually have a tertiary qualification (diploma, undergraduate degree, or postgraduate degree). During assessment any formal qualifications of staff will be considered, together with observation of staff conducting analyses for which they claim competence. Any specific skills are additional to general laboratory skills such as good occupational health and safety practices. In the documentation of the laboratory, job descriptions of key personnel must be maintained. Ongoing training within the organization is also important in maintaining the quality of a laboratory.

New to the 2005 standard is the requirement for an organization to measure the effectiveness of training against predetermined criteria. Sending staff to take courses chosen for the excellence of the cuisine and exotic location might not stand the scrutiny of an assessor when the time comes for renewal of accreditation.

### 9.2.3.3   Subsection 5.3: Accommodation and Environmental Conditions

Sufficient and well-maintained accommodation is essential to a quality laboratory. The standard addresses aspects of security and distinction between areas used for different purposes, although the requirements are not so stringent as those laid out in GLP (see section 9.4.2.3). Good housekeeping is stressed, and assessors usually notice when a laboratory is badly kept, or if there are signs of an unusual and sudden clean up.

### 9.2.3.4   Subsection 5.4: Test and Calibration Methods and Method Validation

The central activity in an analysis usually revolves around a procedure involving the test item, calibrators, and equipment. A measurement is made following a measurement procedure according to one or more measurement principles, using a prescribed measurement method. This statement implies there has been some prior activity to identify the principles and to develop and validate the method. Much of the work will generally have been done by others, but in choosing to use a given method for a given test item, an analyst must use his or her judgment (assuming the client has not specified the method). Whatever method chosen, the client must be informed of it.

For nonstandard methods, the standard 17025 lists 11 items required for the procedure. Methods must be validated (see chapter 7) and measurement uncertainty estimated (see chapter 6). In subsection 5.4, reference is made to the control of data, particularly in relation to electronic or automated equipment.

A large part of the documentation of the laboratory will be of the procedures of the test method and associated sample preparation and handling. However, use of a standard method that is published with sufficient information as to how to carry out the method does not require further documentation.

### 9.2.3.5  Subsection 5.5: Equipment

The requirements for equipment are reasonably obvious: the equipment should be available, appropriate to the method, used by qualified personnel, in good working order, and calibrated. To demonstrate this, proper records of all equipment must be maintained, uniquely identifying the equipment and keeping a history of its maintenance, calibration, and any repairs or modifications. Subsection 5.5 references equipment outside the immediate control of the laboratory, putting the onus on the laboratory for ensuring that this equipment meets the standard. Software that is part of an instrument or that is used to calculate results of a method falls under the requirements of this section. All software should be validated and recorded. In particular, there should be safeguards against unauthorized adjustments to the software (and hardware) that might invalidate results. If equipment has been borrowed or used outside its usual scope, it should be checked before use.

### 9.2.3.6  Subsection 5.6:
### Measurement Traceability

Metrological traceability of measurement results is an important part of ensuring that an analysis is fit for the customer's purpose. The standard was written before a full understanding of the nature of metrological traceability was gained, and there are some statements that conflict with the current VIM definition of metrological traceability (not least the use of the term measurement traceability; metrological traceability is a property of a measurement result, not of a measurement). ISO/IEC 17025 is also written for calibration laboratories whose requirements for metrological traceability are somewhat different from those of a testing laboratory. However, the approach is clear and in the sprit of the modern approach to metrological traceability. All calibrations should be traceable to an appropriate metrological standard. ISO/IEC 17025 goes further than the VIM by specifying that there should be traceability to the international standard (SI) where possible, but where not, the reference used should be clearly described and agreed-upon by all parties concerned. At the end of section 5.6.2.1.2 in the standard is the requirement, where possible, for participation in program of interlaboratory comparisons.

When discussing reference materials and reference standards (standard section 5.6.3), the quantity values carried by reference materials must be traceable to SI units of measurement or to the values carried by certified reference materials. These materials must be stored and handled properly to preserve their reference values. Because the standard covers physical testing and calibration, it mentions "reference standards" that must be traceable and maintained in calibration.

The standard recognizes (section 5.6.1) that the traceability requirements should apply to aspects of the method that have a significant influence on the result of the measurement. For an analytical chemistry laboratory, as well as for the reference materials used for calibrating the response of an instrument, balances will need to be calibrated from time to time, and appropriate certification of the traceability of glassware and thermometers must be available.

### 9.2.3.7  Subsection 5.7: Sampling

Sampling was not part of the original Guide 25, but because the role of the laboratory is seen as more proactive in the present standard, provision has been made for the testing to include acquiring the test material. If the laboratory does sample the test material in the field, then the provisions of the standard are considered to cover this aspect of the test. There must be a sampling plan that is based, whenever reasonable, on appropriate statistical methods. Some sampling will be able to use statistical principles, and there are a number of standards that guide sampling (ISO 1980, 2003). In other circumstances, perhaps at a crime scene, sampling is less methodical[1] and whatever can be sampled is sampled. Any deviations from the designated sampling plan must be documented with the test results and communicated to the appropriate personnel. An accreditation body might place restrictions on the scope of accreditation if it cannot adequately assess the sampling activities of a laboratory.

### 9.2.3.8  Subsection 5.8: Handling of Test
and Calibration Items

The laboratory must have procedures for the proper and safe handling of all test materials, including provisions to ensure the integrity of the test materials. Subsection 5.8 also requires adequate identification of all samples and facilities for tracking samples within the laboratory. Modern laboratory information management systems (LIMS) are used in many laboratories. These systems should be properly validated and have auditing and security facilities. Any concerns about samples (e.g., leaking or damaged containers), should be documented and reported. For forensic samples, chains of custody must be established, and the custody procedures should be documented. Procedures should specify how long after analysis test items should be retained and how they are ultimately disposed.

### 9.2.3.9  Subsection 5.9: Assuring the Quality of
Test and Calibration Results

For a section that embodies the rationale of the standard, subsection 5.9 is remarkably short. The laboratory must have quality control (QC) procedures that are documented, with QC data recorded in such a way that trends are detectable and, where practical, statistical techniques applied. Five QC techniques are offered as a nonexclusive list;

1. Regular use of certified reference materials and/or internal quality control secondary reference materials
2. Participation in interlaboratory comparison or proficiency testing programs
3. Replicate tests using the same or different methods
4. Retesting of retained items;
5. Correlation of results for different characteristics of an item.

There is no mention of control charts or other graphical QC tools (see chapter 4). There is a note that the selected methods should be appropriate for the type and volume of the work undertaken. As part of the emphasis on the overall quality framework, the standard now requires a laboratory to analyze QC data and to implement corrective action if results are outside predefined criteria.

### 9.2.3.10  Subsection 5.10:
Reporting the Results

The outcome of a chemical test is a report containing the results of the analysis and other relevant information. In a recent unpublished survey, one of the most frequent complaints from the clients was the inconsistency and lack of clarity in test reports. Most of the requirements of the standard can be met in a straightforward manner on a single-page form, perhaps on a template that requires the appropriate sections to be completed (see figure 9.2).

There is no standard report form offered because the standard requires it be completed "and in accordance with any specific instructions in the test or calibration methods" (subsection 5.10.1). It is noteworthy that the general notes to subsection 5.10 do allow a simplified report to be issued to internal clients, as long as the full information is readily available in the laboratory. There are 11 items that are required on test reports and calibration certificates (unless there are valid reasons for their exclusion) plus five additional items that cover aspects of interpretation and six items when sampling has also been carried out as part of the test. Increasingly test reports are submitted electronically. This is referred to in the standard (subsection 5.10.7) and is allowed as long as the requirements of the standard are met. Of importance is the integrity of the report and the nature of the signature when the report is sent by e-mail or via the Internet.

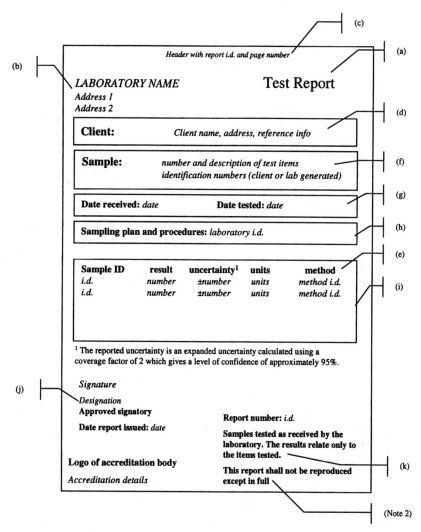

**Figure 9.2.** A template test report that conforms to ISO/IEC 17025. The letters a to i refer to subclauses of 5.10.2 of ISO/IEC 17025.

The ILAC recommends that the logo of the accreditation body not appear on the test report, lest it be taken as a direct endorsement of the test results. However, the marketing value for the accreditation body is as significant as it is to the testing laboratory, and the practice of including the logo is widespread.

Opinions and interpretations are now allowed and covered by the standard (subsection 5.10.5), although not always included in the scope of accreditation. Interpretations are envisaged as including statements about the compliance of the result to a specification, if contractual requirements have been fulfilled, recommendations on the use of the results, and guidance for improvements. Although the ability to offer suitable expert opinion has improved the status of the analyst from mere automaton to a participant in the business at hand, this new-found freedom must be exercised with extreme caution. Obvious caveats are that the analyst must have sufficient expertise to offer the opinion and certainly should not stray outside his or her competence. Analysts should also ensure that they have sufficient knowledge of the problem to make their advice appropriate. The laboratory will be held accountable for any statement made as part of the service paid for, and in this litigious age, loss suffered by the customer on account of the advice will inevitably attract a demand for recompense. Accreditation bodies may not allow opinion to be given on the certified test report if they do not have procedures for assessing opinions offered. In this case the opinion must be presented on a separate sheet, not endorsed by the accreditation body. This is an example of possible mismatch between the standard and the rules of a particular accreditation body.

## 9.3 Accreditation to Good Laboratory Practice

In the mid-1970s, a review by the U.S. Food and Drug Administration (FDA) of toxicology studies on a new drug by a major U.S. pharmaceutical company caused concerns that the studies had not been properly conducted. The FDA requested a "for cause" inspection, which revealed serious discrepancies and evidence of fraud and misinterpreted data. In 1976 the government proposed regulations to prescribe principles of Good Laboratory Practice (GLP) for the nonclinical safety testing of materials in pharmaceutical products, pesticide products, cosmetic products, veterinary drugs, food additives, feed additives, and industrial chemicals. These became effective in 1979, and the U.S. Environmental Protection Agency followed with nearly identical regulations in 1983. This practice was taken up around the world, and in 1981 GLP found a home with the Organization for Economic Co-operation and Development (OECD). The OECD principles were revised and republished in 1998 (OECD 1998). Similar to ISO 17025, but with a less technical and more managerial emphasis, the principles of GLP cover the conditions under which studies are planned, performed, monitored, recorded, archived, and reported. A study that conforms to GLP allows authorities in many countries to recognize test data and so avoids duplicate testing, prevents the emergence of nontariff barriers to trade, and reduces costs for industry and governments. Thus, GLP does for the management of a nonclinical safety testing program what a Key Comparisons mutual recognition arrangement does for more general test results. GLP does not cover the conduct of basic research or the development of analytical methods. A list of the main sections of the OECD-GLP document is given in table 9.4, and a commentary on the sections follows.

Table 9.4. Sections of the OECD-GLP, revised 1998 (OECD 1998)

| Section | Subject |
| --- | --- |
| 1 | Test facility organisation and personnel |
| 2 | Quality assurance programme |
| 3 | Facilities |
| 4 | Apparatus, material and reagents |
| 5 | Test systems |
| 6 | Test and reference items |
| 7 | Standard operating procedures |
| 8 | Performance of the study |
| 9 | Reporting of study results |
| 10 | Storage and retention of records and materials |

### 9.3.1  Section 1: Test Facility Organization and Personnel

Because of its history, GLP emphasizes the need for a single person to be legally responsible for a study. The "study director" is appointed by the management of the organization to oversee the study, to sign off on the plan, and to take responsibility all the way through, until he or she signs and dates the final report, indicating acceptance of responsibility for the validity of the data (subsection 1.2h). An assistant study director is not allowed, but an alternate can be designated to act when the study director is not available. A study director can direct more than one study at a time. The study director is distinct from the principal investigator and other study personnel who must at all times conform to GLP, but are beholden to the study director's overall control.

### 9.3.2  Section 2: Quality Assurance Program

The principles of GLP require an independent quality assurance (QA) program to ensure that the study is being conducted in compliance with GLP. The QA personnel cannot overlap with those of the study because of the potential conflict of interest, but they may be part-time staff if the size of the study does not warrant a full-time QA section. The responsibilities of the QA unit are to maintain copies of plans, standard operating procedures, and in particular the master schedule of the study, and to verify, in writing, that these conform to GLP. The QA unit is responsible for inspections and audits, which must be documented and the results made available to the study director and the principal investigator. The QA unit also signs off on the final report. Any problems discovered or corrective action that is recommended by the unit must be documented and followed up.

### 9.3.3  Section 3: Facilities

The requirements for the facilities for the study are that they be adequate for the needs of the study, particularly with respect to controlling hazards and maintaining the integrity of the test items. There should also be sufficient space for archiving reports and materials and facilities for disposing of wastes.

### 9.3.4  Section 4: Apparatus, Material, and Reagents

Less extensive than ISO/IEC 17025, section 4 of GLP states that all apparatus and materials should be appropriate for the tests, used in accordance with the SOPs, and calibrated, where appropriate, to national or international standards of measurement. Chemicals should be properly labeled, with expiration dates and available material safety data sheets.

### 9.3.5  Section 5: Test Systems

Biological test systems are considered separately from physical and chemical systems, which need only be suitably designed and located, with integrity maintained. For biological tests involving animals, proper records must be maintained, avoiding contamination, and they must be treated humanely. Laboratories conducting field testing in which sprays are used should be careful to avoid cross-contamination.

### 9.3.6  Section 6: Test and Reference Items

All test and reference items should be identified with a Chemical Abstracts Service registry number or other unique code. Where test items are novel compounds supplied by the sponsor of the study, there should be an agreed-upon identification system to allow the item to be tracked through the study. The composition and concentrations of prepared solutions should be known and recorded. A sample for analytical purposes of each batch of test item should be retained for all except short-term studies. This requirement has the potential to overload storage facilities very quickly, and some sensible arrangement must be made to archive samples of importance, while not giving an inspector cause to worry about inappropriate disposal of items.

### 9.3.7  Section 7: Standard Operating Procedures

A standard operating procedure (SOP) is a document that prescribes how to perform an activity, usually as a list of action steps to be carried out serially. The detail in an SOP should be sufficient for a suitably trained and competent person to carry out the actions. It is important that SOPs document what is actually done, not a theoretical procedure. A recent drug case in Australia foundered because the government laboratory was forced to admit that it failed to follow its own SOPs (Hibbert 2003), which could not be properly followed anyway. The laboratory had done what was practical, but the procedures were not in the SOP. An enterprising defense lawyer exploited this fault (*R v Piggott* 2002). Any deviations from the SOP must be approved and documented. Section 7 notes that other documents, books, and manuals can be used as supplements to the SOPs. SOPs can be written for routine inspections, quality assurance and control procedures, analytical methods, sampling, data handling, health and safety precautions, record keeping, storage, handling of test materials, and so on.

Copies of SOPs should be readily available for personnel. There are templates available for writing SOPs that ensure proper history of the documentation, page numbering, authorization, and so on.

### 9.3.8  Section 8: Performance of the Study

The study is performed according to a plan approved (by dated signature) by the study director and verified for GLP compliance by the QA unit. Some countries also require formal approval by the test facility management and the sponsor. The plan should usually contain identification of the study, the test item and reference item, information concerning the sponsor and the test facility, dates, test methods, documents and materials to be retained, and other issues that have been identified.

In the conduct of the study, which must conform entirely to the study plan, all raw data must be available for audit, including computerized records. Proper laboratory notebooks should be used, with numbered pages and signatures of the laboratory staff and witnesses and dates on each page. Any changes should be made so as not to obscure the original entry (i.e., the revision should not be made over the original, correction fluid must not be used, nor heavy crossing out). A line should be lightly drawn through the incorrect entry and the correction made next to it, with an initialed explanation for the change.

### 9.3.9  Section 9: Reporting of Study Results

The outcome of the study is a final report that is approved and signed by the study director, which becomes the official record of the study. Once released, no further reports should be made, except as amendments and addenda to the released report. A final report should have, at a minimum, the following sections:

1. Identification of the study, the test item and reference item
2. Information concerning the sponsor and the test facility
3. Dates

4. Statement of QA procedures and audits and compliance to GLP
5. Description of materials and test methods
6. Results
7. Information of storage of materials, data, and documentation.

### 9.3.10   Section 10: Storage and Retention of Records and Materials

The regulations and laws under which the study is performed might determine the length of time for which materials and documents are stored. Under GLP, the study plan must detail what is to be stored, where it is to be stored, and under what conditions and for how long. To comply with GLP these arrangements must be adequate to allow retrieval and, if necessary, reanalysis of material or data. Access to the archives must be by designated and authorized personnel, and any access must be documented. If the organization conducting the trial ceases to be in business, the archive must be transferred to the sponsoring organization.

An issue that is not addressed, but is of concern is the currency of computer records. Storage from tapes to disks (hard or floppy) to solid-state storage devices presents the problem of reading data more than a couple of years old. It is incumbent on the organization to ensure that any electronic data can be read during the projected life of the archive. This may involve periodically upgrading the storage or storing necessary reading devices with the archive

## References

Hibbert, D B (2003), Scientist vs the law. *Accreditation and Quality Assurance*, 8, 179–83.

ILAC (2001), Guidance for accreditation to ISO/IEC 17025 (Rhodes, Australia: International Laboratory Accreditation Cooperation).

ISO (1980), Water quality—Sampling—Part 1: Guidance on the design of sampling programmes, 5667-1 (Geneva: International Organization for Standardization).

ISO (1993), International vocabulary of basic and general terms in metrology, 2nd ed. (Geneva: International Organization for Standardization).

ISO (1994a), Quality systems—Model for quality assurance in design, development, production, installation and servicing, 9001 (Geneva: International Organization for Standardization).

ISO (1994b), Quality systems—Model for quality assurance in production, installation and servicing, 9002 (Geneva: International Organization for Standardization).

ISO (1999), Chemistry-Layouts for standards—Part 2: Methods of chemical analysis, 78–2 (Geneva: International Organization for Standardization).

ISO (2000), Quality management systems—Requirements, 9001 3rd ed. (Geneva: International Organization for Standardization).

ISO (2003), Statistical aspects of sampling from bulk materials—Part 1: General principles, 11648-1 (Geneva: International Organization for Standardization).

ISO/IEC (1978), General requirements for the competence of calibration and testing laboratories, Guide 25 1st ed. (Geneva: International Organization for Standardization).

ISO/IEC (1999), General requirements for the competence of calibration and testing laboratories, 17025 1st ed. (Geneva: International Organization for Standardization).

ISO/IEC (2004a), Conformity assessment—General requirements for accreditation bodies accrediting conformity assessment bodies, 17011 (Geneva: International Organization for Standardization).

ISO/IEC (2004b), Standardization and related activities—General vocabulary, Guide 2, 8th ed. (Geneva: International Organization for Standardization).

ISO/IEC (2005), General requirements for the competence of calibration and testing laboratories, 17025 2nd ed. (Geneva: International Organization for Standardization).

OECD, Environment Directorate, Chemicals Group and Management Committee (1998), OECD series on principles of good laboratory practice and compliance monitoring, Number 1:OECD principles on Good Laboratory Practice (Paris: Organization for Economic Co-operation and Development).

R v Piggott, Griffiths and Simeon (2002), NSWCCA 218: New South Wales Court of Criminal Appeal, Sydney.

van de Leemput, Peter J H A M (1997), The revision of EN 45001 and ISO/IEC Guide 25, Accreditation and Quality Assurance, 2 (5), 263–4.

# 10

## Conclusions: Bringing It All Together

### 10.1 Introduction

If you have read this book, whether a few pages at a time, by jumping back and forth, or meticulously from beginning to end, the aim of this chapter is to draw together the methods, concepts, and ideas to help you answer the question, how do I make a good analytical measurement? If nothing else, you will have discovered, like the answers to the greater questions of life, that there is not a simple prescription for quality assurance that if followed leads to success. Even knowing if you have the right answer is not always vouchsafed; is customer satisfaction sufficient? Does the continuing solvency of your business say that something must be going well? Does staying within ± $2\sigma$ in interlaboratory studies cause you happiness? The best laboratories do all of this and more. At the heart of a good laboratory is an excellent manager who has recruited good staff, set up a culture of quality, and who understands the science and business of chemical analysis and the requirements of his or her clients. I do not believe laboratories can be run by people with only managerial skills; at some point a chemical analyst is going to have to take responsibility for the product.

In this reprise of the book's contents I revisit the six principles of valid analytical measurement (VAM) so cleverly enunciated by the Laboratory of the Government Chemist. But first some words about clients and samples.

## 10.2  Clients and Samples

### 10.2.1  Clients

As has been stressed throughout this book, many problems can be solved by chemical analysis, and the point of chemical analysis is therefore not to do chemistry for its own sake, but to contribute to the solution of those problems. Clients, or customers as now found in ISO/IEC 17025, come in many shapes and sizes, from people who gladly admit no scientific knowledge at all to fellow professionals who can discuss the analysis as equals. The first kind are more difficult to work with than the latter, although colleagues who meddle are never totally welcome. An apparently simple request to analyze something might require extensive negotiation about exactly what is needed. A written statement of what analytical services are to be provided should be given to the prospective client, and the final report should be judged against this document. After ISO/IEC 17025, the vexing question of reporting measurement uncertainty has arisen. There are still many laboratories that would rather not concern the client with measurement uncertainty, and while they report the reproducibility given in the standard method as their measurement uncertainty to the accrediting body, they do not really engage with the concept. Unfortunately, because measurement uncertainty is at the heart of providing quality results, these laboratories lay themselves open to error. If a client does ask for a statement of measurement uncertainty, which they are entitled to do, then such a laboratory might find it hard to justify their figures. Always stress to the client that the measurement uncertainty is an integral part of a measurement result and completes the "information about the set of quantity values being attributed to a measurand." Measurement uncertainty adds to, rather than detracts from, a result.

### 10.2.2  Samples

A measurement result should identify the measurand as well as give the quantity value and measurement uncertainty. It is the definition of the measurand that is at the heart of the result. Consider the following measurands relating to a 1-L sample from a lake:

1. The amount concentration of chromium in the test sample
2. The amount concentration of chromium(VI) in the test sample
3. The amount concentration of chromium(VI) in the lake
4. The amount concentration of bioavailable chromium(VI) in the lake.

The value of each of these measurands could well be different, with very different measurement uncertainties. If measurement 1 implies total chromium, then it is likely that its value will be of greater magnitude than that of measurement 2. As soon as the result encompasses the larger system, the issue of sampling, only touched upon in this book, becomes important. There is some debate about whether a measurand such as that in item 3 can be measured with metrological traceability because of the sampling step, and some hold that the traceable measurement should stop with the laboratory test sample. Further calculations and estimations might then be done to provide a result for the lake as a whole, but these would be derived results from the information obtained by the analysis. The analyst should only take responsibility for what is in his or her control. With training and experience, the analyst can learn the ins and outs of sampling lakes, and so take sufficient samples to be analyzed. The analyst should not be given a single sample of water labeled "from Lake X" and be expected to return a result reflecting the amount concentration of chromium(VI) in Lake X. Of more dubiety still is measurement 4, where the definition of bioavailability must be carefully stated, and the result is then tied to that definition, in addition to the sampling issues discussed above.

A hypothetical graphical comparison of values and measurement uncertainties for this example is shown in figure 10.1. Note that the usual approach to evaluating measurement uncertainty leads to expanded uncertainties that include zero. Applying the Horwitz estimate for relative standard deviation (see chapter 6) leads to a 100% uncertainty for a relative concentration of about 5 ppt $(1 : 5 \times 10^{-12})$, although such an extrapolation is not advised (see chapter 6). As a final point, when a claim is made about the concentration of a species in something like a lake, the date and time must also be specified, as the measurement will only be valid for the particular moment when the samples were taken. A large part of the uncertainty for results that refer to the lake as a whole will arise from concerns about changes in the composition of the samples when they are taken from the body of the lake.

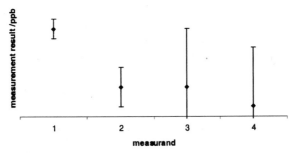

**Figure 10.1.** Four measurement results corresponding to measurands. 1, the amount concentration of chromium in the test sample; 2, the amount concentration of chromium(VI) in the test sample; 3, the amount concentration of chromium(VI) in the lake sampled; and 4, the amount concentration of bioavailable chromium(VI) in the lake sampled.

## 10.3  Valid Analytical Measurement Revisited

So are you now in a position to make valid analytical measurements? Recall the six principles (LGC 2006):

1. Analytical measurements should be made to satisfy an agreed-upon requirement.
2. Analytical measurements should be made using methods and equipment that have been tested to ensure they are fit for purpose.
3. Staff making analytical measurements should be both qualified and competent to undertake the task.
4. There should be a regular independent assessment of the technical performance of a laboratory.
5. Analytical measurements made in one location should be consistent with those elsewhere.
6. Organizations making analytical measurements should have well-defined quality control and quality assurance procedures.

### 10.3.1  Principle 1: Clients

The VAM recommends identifying a "responsible analyst" who can talk to the client and determine what services are needed for the analysis. Aspects that might impinge on the discussion could be the nature of the analyte and matrix, the sampling protocol, the purpose of the analysis, and whether any legal or regulatory limits are involved. Identification of critical issues will help the analyst and the client arrive at the most effective solution in terms of time, cost, and overall quality. The job of the responsible analyst is to interpret the client's desires into parameters of the analysis,—for example, the target measurement uncertainty will be contingent on the confidence the client requires in the result. If there are choices between a quick, but less precise method and a more expensive one, then the likely range of results will need to be assessed. Even if the result must be compared to a limit with a higher degree of precision, it might be decided to use a quicker and cheaper method as a screen and to use the better method for confirming cases for which the result and (higher) uncertainty include the limit. This requires some care on the part of the analyst to educate the client in just what a result and measurement uncertainty conveys.

A benefit of this approach has been to see a more informed body of clients. The old days of "trust me, I'm an analyst" have given way to a partnership in search of appropriate information on which the client can act with confidence.

### 10.3.2  Principle 2: Methods

Chapter 8 considers method validation, the procedures that demonstrate that a method used in a laboratory can achieve a set of predetermined performance criteria. In using standard methods, an organization still needs to show that the method, as practiced in their laboratory with the analysts employed at the time, will give results that follow the criteria. In my opinion this is one of the most often overlooked quality principles. No one expects working laboratories to revalidate entire methods on a regular basis, but simply running a control sample with every batch does not demonstrate the measurement uncertainty often claimed. Older standard methods have reproducibility and repeatability information, and one of these is often passed on as a measurement uncertainty, without consideration of other factors, or even a demonstration that these figures are appropriate for the laboratory. In a recent case in Australia concerned with horse racing, it was emphasized that the measurement uncertainty of a particular test had been agreed-upon by all laboratories in the country, despite evidence that the checks on the ranges of repeat measurements were quite different from one laboratory to another. A point well made by the VAM is that particular care must be taken when making nonroutine measurements. Validation or verification then becomes particularly important.

### 10.3.3  Principle 3: Staff

There are two aspects to having appropriately competent and trained staff. First, the required level of knowledge and training must be determined before employing a person. Whether this implies a postgraduate qualification, a graduate degree, or a technical certificate, the organization needs to be aware of the level of understanding and basic knowledge needed for the particular tasks to be undertaken. A limited understanding of the principles of operation of an instrument or of the basis on which a method is founded can lead to serious errors. A common example seen in students starting their training is using an instrument outside its applicable range. A blessing and a curse of modern instruments is that they are increasingly "idiot proof." An instrument manufacturer will try to sell the idea that the "intelligent" software will prevent a hapless operator from making egregious errors. Unfortunately, artificial intelligence is often more artificial than intelligent, and while some checks along the way are desirable, no instrument can compensate for the truly creative idiot. Paying more for the instrument and less for the operator can be a recipe for disaster.

Having employed a person with the right credentials is only the first step. As with laboratory accreditation, an appropriate qualification is a necessary, but not sufficient, criterion for competence. All staff should undergo careful training when they begin work or when they are asked to undertake new duties. This training should include close scrutiny of their experimental technique, demonstration that they understand quality principles, and verification that they can apply a method to give results consistent with the claimed measurement uncertainty. Good organizations will also make sure that employees have on-going training. Modern analytical chemistry is evolving, with new methods and new applications of methods, and even in the field of metrology in chemistry concepts and methodology are continually under debate. It is ironic that accountants in many countries have to undertake regular training, otherwise they lose their right to practice, while analytical chemists can sit back and enjoy their decades-old qualifications, without having to prove any modern competence.

All training should be documented. Some professional associations have introduced an accredited portfolio for members in which the career of an analyst can be documented and their skills clearly listed. Ideally, performance in any course taken should be assessed. Certificates on a wall for merely showing up for class are not sufficient.

Finally, it should be stressed that well-trained personnel will be better motivated, make better measurements, and will pay for their training many times over.

### 10.3.4   Principle 4: Assessment

Chapter 5 covers the principles and practice of interlaboratory studies, including proficiency testing of laboratories. This principle of the VAM stresses that no matter how careful a laboratory is in making measurements, without comparison with external results, it is possible for a laboratory to have a bias or trend that can go undetected. Even regular analysis of a certified reference material can hide a problem across the complete method of an analysis. Consider a storage procedure that degrades both samples and certified reference materials and calibration standards about equally. Until a proficiency testing sample is analyzed and the result is shown to be very different from other laboratories, such a problem might never be detected (until a client complains). Best of all is regular proficiency testing, when comparison against peer laboratories is made for each sample, but a regular trend can be observed. Is the laboratory getting better? Is its intralaboratory precision acceptable? When there was that one-off bad result, were the causes understood, and did this lead to changes in procedures?

Increasingly a price of accreditation is mandatory participation in proficiency testing schemes. If the scheme test materials really do match the routine analyses performed by the laboratory, proficiency testing gives an excellent assurance to both the laboratory and the client as to the quality of the laboratory's results. If the scheme has assigned quantity values for the test materials sent out, the cohort as a whole can be judged, and biases at the sector level discovered.

### 10.3.5   Principle 5: Comparability

Metrological traceability (chapter 7) is at the heart of comparability of results. As emphasized in chapter 7, very few measurements are made that are not compared with other values. In trade, buyer and seller often analyze a product, the doctor assesses a result against the knowledge of what is healthy, a regulatory agency compares an analytical result against the legal limit, and a manufacturer compares today's results against yesterday's and earlier results. Without confidence in a result, it cannot be compared and so has little use to the customer.

Laboratories must use calibration standards that are metrological traceability to an embodiment of the unit in which they are expressed. This requires use of certified reference materials, usually to make up working standards that calibrate instruments in the laboratory. Implicit is that results will be expressed in common units, and while this is increasingly the case, the example of the Mars Climate Orbiter (chapter 1) should signal those still using versions of imperial units that it might be time to join the rest of the world and use SI units.

The other side of metrological traceability is the proper estimation of measurement uncertainty. Chapter 6 gives the GUM (*Guide to the Expression of Uncertainty in Measurement*) approach, currently espoused by most bodies, and with a philosophical basis in the latest revision of the *International Vocabulary of Basic and General Terms in Metrology* (VIM; Joint Committee for Guides in Metrology 2007). In the third edition of the VIM, many additions and changes have resulted from incorporating the needs of chemical measurements. The draft introduction to the third edition stresses the move away from concepts involving true answers with different kinds of error to the "uncertainty concept" in which a measurement result consisting of a value and uncertainty gives information about the dispersion of values that can be attributed to the measurand after measurement. While this is still filtering down to the practical laboratory, I hope that a better understanding of the basic metrological premises will allow more sensible decisions to be made.

### 10.3.6    Principle 6: Quality Assurance and Quality Control

This final principle in some respects subsumes all the others. A proper QA approach will include communication with the client, valid methods and equipment, trained staff, methods for tracking quality (proficiency testing), and making traceable measurements with appropriate measurement uncertainty. The totality must be stressed. Given conscientious people at different levels in an organization, each contributing to a quality service, an organizational structure and an imprimatur from the wider organization is still needed. In this respect the more laudable statements from quality gurus about a state of mind of quality have some point.

Regular audit and review of the quality system are essential. It must always be in the mind of the QA manager whether the system is doing its job. Having no problems might sound as many warning bells as numerous identified failures. If QC samples (chapter 4) are being tested and flagged when they are outside the 95% confidence intervals, then do 5 out of a 100 actually fail? If not, the tolerances are set too wide. In other words, the system is not necessarily in statistical control. Unfortunately, the job of QC manager is one of constant worry and doubt.

Accreditation to ISO/IEC 17025 or GLP is becoming a standard approach to creating a quality system (chapter 9). Note that accreditation, as such, is not part of the VAM principles, but it is one of alternative quality systems (see chapter 1). Like any quality system, accreditation only fulfills its purpose if implemented properly. Some laboratories have done the minimum to achieve accreditation, and then take the view that having paid for their accreditation payment of the annual fee will then suffice. This practice is evidenced when accredited laboratories perform no better in testing rounds than experienced laboratories without accreditation (see chapter 5).

Modern quality systems are based on a statistical approach to measuring quality (chapters 2, 3, and 4). Chemical analysis has always been at a disadvantage compared to many fields of measurement because of the small number of measurements that can be made. Duplicate measurements are often considered the best that can be achieved, and any more is a luxury. Over longer periods the intralaboratory reproducibility can usually be estimated quite well from the greater set of data available, but the uncertainty associated with small degrees of freedom remains a limiting factor. Multiplying a combined standard uncertainty by 2 to give a 95% confidence interval is often done without a proper regard to the degrees of freedom, which are not always the infinity implied by $k = 2$ (see chapter 6).

## 10.4  Final Words

I have enjoyed writing this book. I have learned much along the way, and although I remain an academic whose day-to-day experience of analytical chemistry is mostly vicarious through my students and others, I have tried to pass on some of the knowledge I have gained. After reading this book, my earnest hope is that a quality assurance manager can be confident enough to implement an appropriate system that suits the laboratory, the organization (the manager's bosses who think they pay for it), and ultimately the organization's clients (who really pay for it).

## References

Joint Committee for Guides in Metrology (2007), International vocabulary of basic and general terms in metrology, 3rd ed. (Geneva: International Organization for Standardization).

LGC (2006), *The VAM Principles*. Valid Analytical Measurement Programme web site. http://www.vam.org.uk/aboutvam/about_principles.asp [accessed 9th October 2006].

# Glossary of Acronyms, Terms, and Abbreviations

| | |
|---|---|
| AAS | Atomic absorption spectrometry |
| AOAC | Association of Official Analytical Chemists |
| ARL | Average run length |
| ASTM | American Society for Testing and Materials |
| BIPM | International Bureau of Weights and Measures |
| CAS | Chemical Abstracts Service |
| CCQM | Consultative Committee on the Amount of Substance |
| CGPM | General Conference on Weights and Measures |
| CIPM | International Conference on Weights and Measrues |
| CITAC | Cooperation in International Traceability in Analytical Chemistry |
| CRM | Certified reference material |
| CV | Coefficient of variation |
| DMADV | Define – measure – analyze – improve – control |
| DMAIC | Define – measure – analyze – design – verify |
| EAL | European Co-operation for Accreditation of Laboratories |
| EURACHEM | A focus for analytical chemistry in Europe |
| EUROMET | Association of European Metrology Institutes |
| GC | Gas chromatography |
| GLP | Good Laboratory Practice |
| GUM | *Guide to the Expression of Uncertainty in Measurement* |
| HPLC | High-performance liquid chromatography |
| ICP-MS | Inductively coupled plasma mass spectrometry |

| | |
|---|---|
| ICP-OES | Inductively coupled plasma optical emission spectroscopy |
| IDMS | Isotope dilution mass spectrometry |
| IEC | International Electrotechnical Commission |
| ILAC | International Laboratory Accreditation Cooperation |
| ISO | International Organization for Standardization |
| IUPAC | International Union of Pure and Applied Chemistry |
| LC | Liquid chromatography |
| LGC | Laboratory of the Government Chemist (U.K.) |
| LIMS | Laboratory information management system |
| LOD | Limit of detection |
| LOQ | Limit of quantitation |
| MAD | Median absolute deviation |
| MRA | Mutual recognition arrangement (agreement) |
| MS | Mass spectrometry |
| MSDS | Material safety data sheet |
| NAMAS | National Measurement Accreditation Service |
| NATA | National Association of Testing Authorities |
| NIST | National Institute for Standards and Technology |
| NMR | Nuclear magnetic resonance |
| OECD | Organization for Economic Co-operation and Development |
| pdf | Probability density function |
| QUAM | *Quantifying Uncertainty in Analytical Measurement* |
| RSD | Relative standard deviation |
| SOP | Standard operating procedure |
| SPC | Statistical process control |
| TBT | Technical barrier to trade |
| VAM | Valid Analytical Measurement |
| VIM | *International Vocabulary of Basic and General Terms in Metrology* |

# Index